I0109671

Pioneering Marine Ecology in the Øresund

Anders Sandøe Ørsted's *De Regionibus Marinis*

Pioneering Marine Ecology in the Øresund

Anders Sandøe Ørsted's *De Regionibus Marinis*

English translation with
introduction and commentary
by Tomas Cedhagen and
Jørgen Ledet Christiansen

AARHUS UNIVERSITY PRESS

Dedication

Tomas Cedhagen dedicates this book to two of his teachers, colleagues and friends at the Tjärnö Marinebiological Laboratory, Strömstad, Sweden: the late honorary doctor Hans G. Hansson and Mr. Bertil Rex, M.Sc.

Jørgen Ledet Christiansen dedicates this book to his daughter Margrete, who is a biologist educated in an academic tradition that can be traced back to, among others, Anders Sandøe Ørsted.

Contents

Acknowledgements

We would like to express our gratitude to the following people (in alphabetical order) for valuable discussions and information during the preparation of this book. Prof Lars Arvidsson, University of Gothenburg, Sweden; Prof Henrik Balslev, Aarhus University; Dr Ole Bennike, GEUS, Denmark; Prof Christer Erséus, University of Gothenburg, Sweden; Dr Jørn Bo Jensen, GEUS, Denmark; Dr Kathe R. Jensen, University of Copenhagen, Denmark; Prof Kerstin Johannesson, University of Gothenburg, Sweden; Prof Ulf Jondelius, Natural History Museum, Stockholm, Sweden; Dr Kennet Lundin, Natural History Museum, Gothenburg, Sweden; Prof Claus Nielsen, University of Copenhagen, Denmark; Dr Ruth Nielsen, University of Copenhagen, Denmark; Ms Anna Persson, Natural History Museum, Stockholm, Sweden; Prof Fredrik Pleijel, University of Gothenburg, Sweden; Prof Marit-Solveig Seidenkrantz, Aarhus University, Denmark; Dr Malin Strand, Swedish University of Agricultural Sciences; Prof Per Sundberg, University of Gothenburg, Sweden; Dr Mikael Thollesson, Uppsala University, Sweden; Dr Anders Warén, Natural History Museum, Stockholm, Sweden. Dr Brian Sorrell, Aarhus University, Denmark, kindly revised the rendering into English.

Introduction

Marine biology studies

Anders Sandøe Ørsted was born in Rudkøbing in Southern Denmark on 21 June 1816 to the grocer Jacob Albert Ørsted (1780–1829) and his wife Petronelle Catherine Bang (1781–1845). Today, Ørsted's name is usually spelled Ørsted; however, in his time the German and Swedish letter Ö was generally used. In 1820, he went to live in the house of his uncle with the same name, Anders Sandøe Ørsted (1778–1869), who took care of him as if he were his own child.

In 1835 he graduated from Borgerdydskolen in Copenhagen. In 1836 he passed the *Examen philosophicum*, the exam that all Danish university students from 1675 until 1971 had to pass before they could go on to actual academic studies. He then, in 1837, became a natural history teacher at Borgerdydskolen, and in 1839 at the Metropolitan School. At the beginning of the 1840s, he developed a lively career in zoology. He won the University's Gold Medal in 1841 for an article on Denmark's annelids. In 1843 and 1844 he published some works on Greenland's annelids and flatworms, which among other things enriched the knowledge of Nordic fauna very considerably, constituting a turning point in flatworm research. On 29 April 1844 he defended his master's dissertation, which carried the title *De Regionibus Marinis*. In 1845 he began a journey of more than three years to the Caribbean (Danish West Indies and Jamaica) and Central America (Nicaragua and Costa Rica) with zoological studies in mind. For this purpose, he applied to the Royal Danish Academy of Sciences for funds to produce equipment with which he could collect animals down to a depth of 1,000 fathoms. He also brought home rich collections and many color drawings of marine invertebrates made shortly after they had been caught. These collections were processed only in small part by himself, and although his debut as a zoologist was quite successful, he soon abandoned zoological studies, and after 1850 barely continued them. The annelids he had collected were turned over to Adolf Eduard Grube (1812–1880) in Königsberg.

Ørsted set out as a zoologist but returned home as a botanist after his encounters with Central America's lush tropical vegetation, becoming professor ordinarius of botany in Copenhagen from 1862 until his death. Primarily, he worked with Central American plants and mycology during this period. He died of dysentery on 3 September 1872 (Warming 1881; 1905; Meisen 1932; Spärck 1933; 1962; Dahl 1941; Wolff 1967; 1979; Sandbeck 2007; Hansson 2011).

Two important biographies of Ørsted exist. Johannes Eugenius Bülow Warming (1841–1924) was a student of Ørsted's and wrote a detailed biography of him in 1905. Christensen (1924–1926) gave a detailed review of Ørsted's work which made important comments on several of his publications and included a biography.

De Regionibus Marinis

Anders Sandøe Ørsted strove for a synoptic view in his studies of the natural conditions of the sea, taking his point of departure in the Øresund – a small strait between Denmark and Sweden. His dissertation is pioneering work as it marks the beginning of benthic ecology. Distributional patterns of the living organisms were connected to environmental factors like geological background and sediment types, and hydrographical factors such as depth, current and light (Warming 1905; Wolff 1967; 1979).

Ørsted defended his thesis *De Regionibus Marinis* for the award of the master's degree at a disputation in Copenhagen on 29 April 1844. The respondent was the military surgeon and botanist Emil Petit (1817–1893). Zoologists and medical doctors shared the same basic training until the twentieth century. The branch of science we now call botany was also often referred to as '*Materia medica*' in the 19th century. The opponent must therefore have been well qualified for his role as Ørsted's thesis covered zoology, botany and chemistry. As Ørsted did not have an "Embedseksamen" (equivalent to a Bachelor of Science), he requested dispensation to defend his thesis. This was granted on the grounds that Ørsted had won two gold medals and had published several scientific articles that had been recognized. Two other people requested similar exemptions in the same year, which were also granted. These three exemptions were attacked in the newspaper Fædrelandet, but Ørsted was attacked much more strongly than the other two. The reason behind this was that there was an indirect attempt to damage Ørsted's uncle (of the same name), who was the Danish Prime Minister. Ørsted himself did not take part in this polemic, but it later had negative consequences for him (Christensen 1924–1926).

Ørsted was influenced to conduct this study by a number of earlier scientists. He makes several references to the chemist and geologist in Copenhagen, Johan Georg Forchhammer (1794–1865). He must have been a mentor for Ørsted and introduced him to several aspects of geology. Ørsted refers to Forchhammer in the most respectful way but dares to express some deviating opinions based on his own observations.

In the geological chapter, Ørsted also mentions the Cimbrian flood, also called the Cymbrian flood by others, as a causative factor behind geological structures in the Sound. According to Roman authors, the Cimbrians were a people living in Jutland but forced to migrate by severe flooding in the period 120–114 BC. They went south and, together with other German tribes, the Ambrones and the Teutons, they came into conflict with the Romans between 113–101 BC during what became known as the Cimbrian Wars (Roller 2014). C.G.J. Petersen discovered a little horse mussel in Kattegat, which he misidentified as *Modiolus adriaticus* Lamarck, 1819. In fact, that species dispersed from the south during the warm Stone Age period and survived in the Skagerrak-Kattegat area where it evolved into an endemic species that was later described under the name *Modiolus cimbricus* Ockelmann & Cedhagen, 2019 in order to commemorate the Cimbrian tribe. In fact, the Ice Age with its glacifluvial remains was not discovered and understood until a few years after Ørsted's death, when the Swedish geologist and zoologist Otto Martin Torell (1828–1900) found ice grooves and an erratic block of Kinne diabase outside Berlin (Frängsmyr 1976; Nejle 1998).

Ørsted also refers to the book *Principles of Geology* (1837) by Charles Lyell (1797–1875). This is a book of particular importance. Previously, geological processes were usually interpreted in the light of the Bible, and often believed to be catastrophic processes (*e.g.*, Swedenborg 1719; Linnaeus 1747; Cuvier 1839). On the contrary, Lyell argued that catastrophes in geology are rare events, and that the world we see around us is primarily the result of the same, often slow, geological processes that we experience today, but over enormously long periods of time. Lyell travelled extensively and gave numerous examples of observations and conclusions based on these journeys. In addition to being a book on geology, it could also be called a handbook of modern scientific reasoning and inference – observe and make conclusions. Captain Robert FitzRoy (1805–1865) gave this book to the young Charles Darwin (1809–1882) when they embarked on their five-year expedition on the HMS Beagle. This book was eye-opening for Darwin, in forming his formulation of the theory of evolution. It probably had the same enlightening effect on Ørsted.

Among botanists, he mentioned the early authors Joseph Pitton de Tournefort (1656–1708) and Carolus Linnaeus (1707–1778) as doing pioneering work in plant systematics and geography. Ørsted respected Alexander von Humboldt (1769–1859) who did extensive research on quantitative plant geography that laid the foundation for the research field of biogeography. This field of science was solidly established by his publications "Essai sur la géographie des plantes" and "Tableau physique des régions équinoxiales" (von Humboldt & Bonpland 1805; 1807). Other scientists of importance for him were Göran Wahlenberg (1780–1851), Robert Brown (1773–1858) and Augustin Pyramus de Candolle (1778–1841), the last of whom coined the term 'taxonomy'. It is surprising that he did not refer to Jean Vincent Félix Lamouroux (1779–1825) who in 1813 had recognized the green, brown and red algae. They form the basis for Ørsted's division of the sea into regions (zones). Frederik Michael Liebmann (1813–1856) and Hans Christian Lyngbye (1782–1837) as well as Jacob Georg Agardh (1813–1901) gained some understanding of such a zonation for the algae, and it is also this vegetation that forms the basis for Ørsted's division (Agardh 1835).

Joachim Frederik Schouw (1789–1852) was a professor of botany at University of Copenhagen and a teacher of Ørsted's. He was a pioneer in plant geography and the author of one of the very earliest books in this field – *Grundtræk til en almindelig Plantegeographie* [Fundaments of a general Phytogeography] in 1822. Ørsted cited his book and gave him credit as one of the founders of this research field. The term 'plant geography', as used by Ørsted, includes phytogeography, but also other aspects for which we today use the term 'ecology' – a word constructed later by Ernst Heinrich Philipp August Haeckel (1834–1919). Finally, Ørsted (1843) gave much credit to the Danish marine biologists Otto Fabricius (1744–1822) and Otto Friedrich Müller (1730–1784) (Kornerup et al. 1923).

Ørsted's master's thesis is a natural history monograph of the Sound, in which he seeks to find the chemical and physical conditions that determine the distribution of plants and animals in the oceans (Warming 1905; Wolff 1967; 1979). What Ørsted called regions are today known as zones (Lewis 1964). However, Ørsted's new approach was that he tried to understand the causes behind the patterns he observed and therefore gives an overview of the chemical nature and movements of water and of the character of the bottom of the sea.

Ørsted separated the vegetation into four algal zones from the shoreline downwards. Each was dominated by the blue-green, green, brown, and red algae, respectively. The distribution, or zonation, of the different groups of algae is primarily determined by the difference in depth to which the different colors of light can penetrate. Different groups of algae have a greater or lesser ability to utilize the different colors of light (Hansson 2011; Warming 1905; Wolff 1967). In addition to the light, Ørsted pointed out other factors such as the substrate.

He also argued that the animals occurred in regions (zones) that were to a great extent determined by the plant zones (regions), but other factors – depth, temperature, nature of the bottom, currents, etc. – were likewise important. In his "Historical introduction" he refers to several scientists who have worked with animal geography (zoogeography) but mainly on terrestrial animals. He also clearly pointed out the geography of marine animals as an under-studied research field and gave only a few examples.

The equipment Ørsted used while researching his thesis was called Ball's dredge. It is made of iron, resembles an oyster dredge and is very heavy. During sampling it is towed by a rope from a small boat propelled by oars or sail. After dredging, it was heaved back into the boat, full of organisms and sediment, by muscle power alone. A very large number of samples must have been collected in order to get as good a picture of the organisms as Ørsted created. We know from experience that it is extremely tiring to operate similar equipment, so Ørsted must have been very fit after completion of the fieldwork.

Some of Ørsted's results were incorrect, but the basic idea was sound and has given rise to other studies and a deeper understanding still relevant today (Hansson 2011; Wolff 1967).

The reception of De Regionibus Marinis

During Ørsted's time, very few marine biologists were interested in what we today call marine ecology. However, one of the few who did share this interest was the British naturalist Edward Forbes (1815–1854). He conducted sampling using dredges in the Mediterranean as well as in British waters and was one of the first to work with quantitative methods. In particular, he studied diversity and biomass and how they gradually changed with depth and distance from land. Ørsted was unaware of his investigations and achieved his results on his own. Forbes cited Ørsted in his posthumously published book *The Natural History of the European Seas* (1859) and gave an extensive presentation of Ørsted's results. His work, however, was of little importance compared with the influence that Forbes came to have in the following decades, as it was his theory of the division of marine animals by depth that was the predominant one.

Sven Lovén (1809–1895) gave a summary of Ørsted's principal results in 1852. In particular, he developed aspects around the importance and changing character of light with increasing depth and how it affects the animals living there. This discussion is extended with several examples from the local region as well as from distant parts of the world. At that time, based on the colors of algae and animals, it was believed that blue light was absorbed in the shallowest water and that red light penetrated to the greatest depth. The order of the colors followed the spectrum. Today, we know that this is wrong (Denny 2008): the long-wavelength light (red) is absorbed closest to the surface and the most energy-rich and short-wavelength light (blue) penetrates to the greatest depth. There was no method to investigate or understand these aspects of light absorption in Ørsted's and Lovén's time. However, Ørsted's basic idea was right. Engelmann (1884), Engelmann & Gaidukov (1902), and Gaidukov (1904) later formulated the hypothesis of "complementary chromatic adaptation". It means that effective absorption of light requires accessory pigments that absorb the color of the incident light; under white light, the pigment appears to have the color complementary to the wavelength of light that it absorbs. Consequently, red algae occur deepest where the incident light is mainly green. Green algae occur closest to the surface where they most effectively utilize the red light. Brown algae use red and blue light most effectively, but this ability is extended to the blue-green parts of the spectrum. Consequently, they have an intermediate position regarding depth distribution (Kinne 1970; Denny 2008) and can therefore occur also at great depths.

Frans Reinhold Kjellman (1846–1907) cited Ørsted in 1878 and wrote about differences in the algal zones compared with the Swedish west coast. He seemed to appreciate Ørsted's book and concluded, for example, that the borders between the zones were less distinct in Sweden than reported from the Sound by Ørsted.

Most authors agree with Ørsted in his general description of algal zones, but they mostly hesitate about the existence of distinct borders between these zones. Axel Johan Einar Lönnberg (1865–1942) conducted studies similar to Ørsted's that were published in 1898, 1899 and 1903. He called Ørsted's book excellent but expressed an alternative idea to Ørsted's zones. One should rather distinguish a beach region, an eelgrass (*Zostera marina*) region and an algal region, but these in turn should be divided into a greater or lesser number of sub-regions or formations after the leading and dominant plants in such a way as Kjellman did in the treatment of Skagerack's algal vegetation (Lönnberg 1898). He also gave an extensive discussion about the many cold-water or Arctic species that are restricted to, or particularly abundant in, the Sound region. Björck (1913, 1915) also appreciated Ørsted's results very much and, like Lönnberg, mentioned the large number of Arctic species found by him.

Several other scientists doing marine biological research cite Ørsted (1844) but most of them seem to be quite uninterested in the ecological aspects. They focus on the floristic and faunistic results (Meyer & Möbius 1865a, b; Jeffreys 1869). Up to about 1880, then, the efforts of marine research had, as in other countries, been directed chiefly towards registering the species of flora and fauna. This is also true for the book *Om Limfjordens tidligere og nuværende marine fauna, med særlig hensyn til bløddyrfaunaen* about the fauna of Limfjorden, published in 1884 by Jonas Collin (1840–1905). Like Ørsted, he attempted to give a general description of a marine region, Limfjorden, but the ecological or natural history information is limited to comments under various species in this primarily faunistic overview. Ørsted's book about the Sound is really the only publication that moves beyond this line.

At the same time as the older period in Danish marine research eventually died out with Collins Limfjord's book, a new era dawned (Nordenskiöld 1882), the era of marine biology, in which we are still living, and which was already in full development abroad. It is fitting that the man who ushered in this epoch and became its leading figure over a long period, Carl Georg Johannes Petersen (1860–1928), adopted, in one of his first writings, a quote from Ørsted as a motto: "I believe that numerous organisms in the sea are not widespread by chance, but according to certain laws." Petersen's entire enterprise can be said to be an attempt to find the definite laws at whose existence Ørsted had guessed, just as it became a realisation of the program that Georg Winther had drawn up for the marine research of the future (Dahl 1941).

The marine biology tradition from Ørsted until today

During the last decades of the 1800s, a new era dawned. As mentioned, one of the prime movers of this new era was C.G. Johannes Petersen. He is often considered as the founder of modern fisheries research (Dahl 1941). Petersen extended Ørsted's investigations to the entire Kattegat-Skagerrak and further to the North Sea and coasts near Norway and Sweden (Petersen 1888; 1893). Based on these and later investigations (Petersen & Boysen Jensen 1911; Petersen 1913; 1918) he was the first to describe benthic communities of marine invertebrates. For this purpose, he constructed a device for quantitative benthic sampling, the Petersen grab, which is still in use today (Thorson 1938).

In 1889, Petersen became a co-founder and first director of the Danish Biological Station. For many years this station was an old naval transport vessel. It was towed to a new position each spring and remained there over the summer. The extensive results were published in "Reports of the Danish Biological Station". From 1934, the Station was located in Charlottenlunds Castle and was in 1952 merged with the Commission for Fishery and Marine Investigations to form the Danish Fisheries and Marine Research Institute. A special department deals with plankton studies (Blegvad 1943).

Petersen in his work felt a certain connection to Ørsted. He was not primarily interested in details or peculiarities, but in the whole, in general factors, as Ørsted was. Gunnar Axel Wright Thorson (1906–1971) was Petersen's last student. The proud young student found a very rare mollusc and showed Petersen a specimen. Petersen replied to Thorson, "Throw it away. It does not say anything to us" (Jørgen Hylleberg, Thorson's last student, pers. comm.).

The tradition of marine research begun by A.S. Ørsted and continued by Petersen thrives today at all Danish universities, in governmental, regional and municipal research institutes, and in consultant companies. Gunnar Thorson founded the Marine Biological Laboratory in Helsingør, a part of Copenhagen University, and there was an extensive cooperation with other Danish and international institutes, particularly the Zoological Museum in Copenhagen. There, among many other things, he continued the research on benthic communities and extended it to other parts of the world.

Petersen inspired corresponding research on benthic communities in other countries, for example in Sweden by Jägerskiöld (1971), Molander (1928; 1962), Gislén (1929; 1930) Lindroth (1935; 1971) and Göransson (2002). These extensive studies have later been followed up by, among others, Pearson et al. (1985), Rosenberg et al. (1986) and Svane & Gröndahl (1988). Gislén's investigation on "marine sociology" (1930) was even cited in the novel *Log from the Sea of Cortez* by the Nobel Prize laureate John Steinbeck.

The 100-year jubilee of Ørsted's book *De Regionibus Marinis* was commemorated by articles written by two prominent marine biologists, Gunnar Thorson (1944) and in 1945 by Hans Olof Brattström (1908–2000). Brattström in particular gives an extensive review of all investigations in the Sound.

In translating Ørsted's book, we have updated his species nomenclature according to the World Register of Marine Species (WoRMS, marinespecies.org) to make it more easily readable by modern biologists. A complete checklist with all taxa from Ørsted's book and the corresponding updated nomenclature is included here (see List of taxa).

Changes in flora and fauna since the time of Ørsted

Some organisms reported by Ørsted have today disappeared from the Sound and Kattegat area, and other organisms have arrived. There are several possible reasons for this.

1. Ørsted conducted his investigation during the cold period called The Little Ice Age (c. 1300 – before 1900). Later scientists, from Lönnberg and onwards, started investigations at the warmer period after the Little Ice Age. Some examples of a transition from a cold-water fauna to a slightly more temperate one are given below.
   a. The nudibranch gastropod *Hero formosa* was reported by Ørsted. It is a cold-water species but has not been found in Denmark after that.
   b. The gastropod *Typhlomangelia nivalis* was frequently collected alive in the Kattegat-Skagerrak area during the 1800s, but not later. Remaining empty shells of this species could be collected in this area until around 1960, but since then they have disappeared (collections in the natural history museums in Gothenburg and Stockholm), either by being passively buried in the sediment by, e.g., trawling or sediment reworking, or by erosion.

c. The gastropod *Trivia arctica* – the only cowrie snail in this region – was often collected in Bohuslän during the 1800s and jars with dozens of individuals, where each jar is the result of a few weeks' field-work, exist in the collection of the Natural History Museum in Stockholm. Since the end of the twentieth century, it is rarely seen. During fieldwork for monitoring or intense sampling for teaching in the field, only single individuals are found nowadays. In some years, the species is not found at all.
d. The circum-Arctic cold-water seastar *Pontaster tenuispinus* was collected in Bohuslän during the 1800s but has not been found subsequently (see also Dahl & Hanström 1972). The species was described based on samples collected in Norway and Sweden (Düben & Koren 1846). Samples are stored in the collections of the Natural History Museum in Stockholm and type material in the Zoological Museum, University of Lund, Sweden.

2. The Kattegat and Sound area is one of the areas most affected by heavy ship traffic in the world. Large commercial ships are even today painted with tributyl tin (TBT). When dissolved in seawater, it causes a phenomenon called imposex. It means that female gastropods develop male secondary sex organs such as a penis and vas deferens, which makes the gastropods sterile. This toxicity, combined with the

low water exchange rate and internal re-circulation of the water mass in Kattegat, could increase the prevalence of imposex and cause the observed reduction of many gastropod species in the Kattegat-Skagerrak area (Strand & Jacobsen 2002).

3. Coastal eutrophication is an important factor in the region. It means that the sea becomes progressively enriched with minerals and nutrients that increases and alters the primary and secondary production thus causing a different composition of biota. During Ørsted's time before the sea was eutrophicated, the seawater was generally clearer than today. Light could therefore penetrate deeper so that the benthic vegetation could live deeper than today. Ørsted reported the lower limit for vegetation to be 36 m and Lönnberg (1898) wrote that "kelp" occurs down to a depth of at least 25 metres. A study of brown algae in the Baltic Sea in 1984 (Kautsky et al. 1986) showed that their depth distribution had decreased since the 1940s (Waern 1952) because of reduced water transparency.

4. Bottom trawling is one of the most violent of the large-scale human activities that affects and destroys the sea bottom (Olsgard et al. 2002). It was formerly also practised in the Sound but has been forbidden since 1932 (Petersen et al. 2018).

Plankton algae of the oceans

Ørsted's greatness is not limited to benthic ecology. During the journey to America, Ørsted discovered that planktonic microalgae, ubiquitous in the sea, form the basis for the marine food web. Based on this pioneering work, he published a small treatise in Danish on "a hitherto unknown, common distribution of microscopic plants in the world ocean" in 1849. He had found the oceans to be full of small algae and drew the conclusion from it that the same must be the case everywhere. This discovery was the solution to the riddle as to where the animals of the sea ultimately got their food from. The result was received with skepticism by his colleagues and has mostly been forgotten or neglected. However, this is the central and most important of all results ever published in plankton research and marine ecology (Warming 1905; Spärck 1933; Dahl 1941; Wolff 1967; 1979; Hansson 2011). John Vaughan Thompson (1779–1847) is credited for being the first to use a plankton net in 1816 (Campbell 1989). However, it had a coarser mesh size than that used by Ørsted, and so Thompson was therefore unable to discover what Ørsted did.

The term "plankton", a Greek word for the living organisms drifting in the sea, was introduced about 40 years after Ørsted's discovery by Victor Hensen (1835–1924) in Kiel. Hensen, not Ørsted, is generally, but erroneously regarded as the pioneer of this concept in plankton research and realized the significance of phytoplankton to the food webs in the sea. However, Hensen's systematic plankton research ensured that the significance of phytoplankton for the primary production in the sea became generally recognized (Meisen 1932; Dahl 1941; Wolff 1967). The Norwegian phycologist Johan Nordahl Fischer Wille (1858–1924) said in 1904 that Ørsted's views "were astonishingly astute, at that time almost prophetic" (Warming 1905).

Ørsted recorded numerous planktonic algae wherever he searched for them in the Atlantic Ocean, and in the easternmost parts of the Pacific Ocean that he also visited during his journey to the west coasts of Nicaragua and Costa Rica. In his article, he extended his observations to be valid also for all other oceans by citing other scientists and seafarers. Most of these older observations probably represent planktonic algae. However, the older reports of "marine sawdust" cited in his article are most probably not unicellular algae. The marine sawdust can form enormous slicks on the surface of the sea and was particularly frequently reported from the waters of New Holland (= Australia). It is a yellow to orange grainy mass occurring during certain periods. "Marine sawdust" is most probably the result of mass spawning of hermatypic corals. This phenomenon was first reported and understood during the 1980s and has been observed in all oceans where tropical coral reefs occur (Harrison et al. 1984).

Despite its important results, Ørsted's 1849 article is very infrequently cited. We include an English translation of this article as an Appendix to this work.

Ørsted's reputation

It is striking, and surprising, that Ørsted is not mentioned in later larger monographs on the history of biology such as those of Nordenskiöld (1935), Söderqvist (1986), Bonde et al. (1996), Worster (1996) or Uddenberg (2004); or in monographs in benthic ecology (Lewis 1964; Moore & Seed 1986) or plankton ecology (Hardy 1965; Kiørboe 2008). There could be various reasons for that. One reason for being forgotten today is that Ørsted's dissertation was written in Latin. This was the common language for scientific publication and communication until the Congress of Vienna in 1848. This Congress marked the reaction against the Napoleonic era and greatly boosted nationalism. One consequence of this was that numerous scientists began to publish their findings in their various native languages such as Danish, English, French, German, Norwegian, Swedish and many others. Consequently, Ørsted's dissertation was one of the last ones published in Latin at the University of Copenhagen. The first dissertation in zoology in Danish was printed and defended in 1857 by Christian Frederik Lütken (1827–1901).

In the field of marine research, Ørsted became the first Danish ecologist, and one of the earliest in the world. His original and independent research lead to two central discoveries in marine ecology: the zonation of benthic organisms in the sea and its relation to causative factors, and the role of phytoplankton as a main primary producer in the sea. He therefore deserves to be regarded as one of the fathers of modern marine ecological research. Several scientists have appreciated the work of Ørsted and honored him by naming a multitude of organisms after him (Table 1).

References

Agardh, J.G. 1836. *Novitae florae Sveciae ex Algarum familia, quas in itineribus ad oras occidentales Sveciae annis 1832–35 collegit et cum observationibus diagnosticis et geographicis*. Lund: Berling.

Björck, W. 1913. 'Biologisch-Faunistische Untersuchungen aus dem Öresund. I. Pantopoda, Mysidacea und Decapoda.' *Lunds Universitets Årsskrift* N.F. Afd. 2. 9(17): 1–39, tab., map.

Björck, W. 1915. 'Biologisk-faunistiska undersökningar af Öresund. II. Crustacea Malacostraca och Pantopoda.' *Lunds Universitets Årsskrift* N.F. Afd. 2. 11(7): 1–98, map.

Blegvad, H. 1943. 'Dansk Biologisk Station gennem 50 aar 1889–1939.' *Beretning fra Den Danske Biologiske Station* 45: 1–67.

Bonde, N., Hoffmeyer, J. & Stangerup, H. (eds.) 1996. *Naturens historiefortællere*. 2 vols. Copenhagen: G.E.C. Gads Forlag.

Brattström, H. 1945. 'Smärre undersökningar över Öresund 11. "De Regionibus Marinis". 100 års zoologisk öresundsforskning.' *Kungliga Fysiografiska Sällskapets i Lund Förhandlingar* 13(16): 1–10.

Campbell, H. 1989. 'John Vaughan Thompson, F.L.S.' *Proceedings of the Linnean Society of New South Wales* 111(2): 45–64.

Christensen, C. 1926. *Den Danske Botaniks Historie med tilhørende Bibliografi*. 3 vols. Copenhagen: H. Hagerups Forlag.

Collin, J. 1884. *Om Limfjordens tidligere og nuværende marine fauna, med særligt hensyn til bløddyrsfaunaen*. Copenhagen: Gyldendalske Boghandels Forlag.

Cuvier, G.L.C.F.D. 1821. *Ideer om Orsakerna till Jordytans närvarande form och om de Revolutioner den undergått*. Stockholm: Henrik A. Nordström.

Dahl, E. & Hanström, B. 1972. *Djurens värld, vol. 4, Ryggradslösa djur*. Malmö: Förlagshuset Norden AB.

Dahl, S. 1941. *Den danske plante- og dyreverdens udforskning*. Copenhagen: G.E.C. Gad.

Denny, M. 2008. *How the Ocean Works. An Introduction to Oceanography*. Princeton and Oxford: Princeton University Press.

von Düben, M.W. & Koren, J. 1846. 'Öfversigt af Skandinaviens Echinodermer [Overview of Scandinavian Echinodermata].' *Kongliga Svenska Vetenskaps-Akademiens Handlingar* 1844: 229–328, pls. VI–XI.

Engelmann, T.W. 1884. 'Untersuchungen über die quantitativen Beziehungen zwischen Absorption des Lichtes und Assimilation in Pflanzenzellen.' *Botanische Zeitschrift* 42: 83–94.

Engelmann, T.W. & Gaidukov, N.I. 1902. 'Über experimentelle Erzeugung zweckmässiger Änderungen der Färbung pflanzlichen Chromophylls durch farbiges Licht.' *Archiv für Anatomie und Physiologie* (Abteilung für Physiologie, Supplementband): 333–35.

Forbes, E. 1859. *The Natural History of the European Seas*. London: John van Voorst.

Frängsmyr, T. 1976. *Upptäckten av istiden: studier i den moderna geologins framväxt*. Stockholm: Almqvist & Wiksell.

Gaidukov, N. 1904. 'Zur Farbanalyse der Algen.' *Berichte der Deutschen Botanischen Gesellschaft* 22: 23–29.

Gislén, T. 1929. 'Epibioses of the Gullmar Fjord I. A study in marine sociology.' *Skriftserie utgiven av K. Svenska Vetenskapsakademien* 3: 1–123.

Gislén, T. 1930. 'Epibioses of the Gullmar Fjord II. A study in marine sociology.' *Skriftserie utgiven av K. Svenska Vetenskaps-akademien* 4: 1–380.

Göransson, P. 2002. 'Petersen's benthic macrofauna stations revisited in the Öresund area (southern Sweden) and species composition in the 1990s – signs of decreased biological variation.' *Sarsia* 87: 263–80.

Hansson, H.G. 2011. 'BEMON. Biographical Etymology of Marine Organism Names.' http://www.marbef.org/modules/imis/?name=People&module=person&persid=3095. (Accessed 12 September 2012).

Hardy, A. 1965. *The Open Sea: Its Natural History*, 2 parts. Boston: Houghton Mifflin Co.

Harrison, P.L., Babcock, R.C., Bull, G.D., Oliver, J.K., Wallace, C.C. & Willis, B.L. 1984. 'Mass Spawning in Tropical Reef Corals.' *Science* 223(4641): 1186–89.

von Humboldt, A. & Bonpland, A. 1805. *Essai sur la géographie des plantes : accompagné d'un tableau physique des régions équinoxiales, fondé sur des mesures exécutées, depuis le dixième degré de latitude boréale jusqu'au dixième degré de latitude australe, pendant les années 1799, 1800, 1801, 1802 et 1803*. Paris: Chez Levrault, Schoell et compagnie, libraires, XIII.

von Humboldt, A. & Bonpland, A. 1807. *Voyage de Humboldt et Bonpland (Première partie. Physique Générale, et relation historique du voyage. Premier Volume, Contenant Essai sur la Géographie des plantes, accompagné d'un Tableau physique des régions équinoxiales, et servant d'introduction à l'Ouvrage)*. Paris: Chez Fr. Schoell.

Jeffreys, J.G. 1869. *British conchology, or an account of the Mollusca which now inhabit the British Isles and the surrounding seas*, vol. 5. London: John van Voorst.

Jägerskiöld, L.A. 1971. 'A Survey of the Marine Benthonic Macro-Fauna Along the Swedish West Coast 1921–1938.' *Acta Regiae Societatis Scientiarum et Litterarum Gothoburgensis. Zoologica* 6: 1–146+map.

Kautsky, N., Kautsky, H., Kautsky, U. & Waern, M. 1986. 'Decreased depth penetration of *Fucus* (L.) since the 1940's indicates eutrophication of the Baltic Sea.' *Marine Ecology – Progress Series* 28: 1–8.

Kinne, O. (ed.) 1970. *Marine Ecology. A Comprehensive, Integrated Treatise on Life in Oceans and Coastal Waters*, vol. 1. Environmental Factors, Part 1. London: Wiley-Interscience.

Kiørboe, T. 2008. *A mechanistic approach to plankton ecology*. Princeton and Oxford: Princeton University Press.

Kjellman, F.R. 1878. 'Ueber die Algenregionen und Algenformationen im östlichen Skagerrack, nebst einigen Bemerkungen über das Verhältniss der Bohuslänschen Meeresalgenvegetation zu der Norrwegischen.' *Bihang till Kongliga Svenska Vetenskaps-Akademiens Handlingar* 5: 1–36+pl.

Kornerup, B., Schultz-Lorentzen, C.W., Jensen, A.S. & Fabricius, O. 1923. 'Biskop Dr. Theol. Otto Fabricius. Et mindeskrift i hundredeaaret for hans død.' *Meddelelser om Grønland* LXII: 217–400.

Lamouroux, J.V.F. 1813. 'Essai sur les genres de la famille des thalassiophytes non articulées.' *Annales du Muséum d'Histoire Naturelle [Paris]* 20: 21–47, 115–39, 267–93.

Lewis, J.R. 1964. *The Ecology of Rocky Shores*. London: The English Universities Press.

Lindroth, A. 1935. 'Die Associationen der marinen Weichböden. Eine Kritik auf Grund der Untersuchungen im Gullmars-Fjord, Westschweden.' *Zoologiska Bidrag från Uppsala* 15: 331–68.

Lindroth, A. 1971. 'On biocoenoses; coenotypes and coenoclines.' *Thalassia Jugoslavica* 7(1): 185–94.

Linnaeus, C. 1747. *Wästgöta-Resa, På Riksens Högloflige Ständers Befallning Förrättad år 1746. Med Anmärkningar uti Oeconomien, Naturkunnogheten, Antiquiteter, In vånarnes Seder och Lefnads-Sätt, Med Tilhörige Figurer*. Stockholm: Lars Salvius.

Lovén, S. 1852. *Berättelse om framstegen i molluskernas, crustaceernas och de lägre skelettlösa djurens naturhistoria under åren 1845–1849*. Stockholm: P.A. Norsted & Söner.

Lyell, C. 1837. *Principles of Geology*, vol. 3, 5th ed. London: John Murray.

Lönnberg, E. 1898. 'Undersökningar rörande Öresunds djurlif.' *Meddelanden från Kongliga Landtbruksstyrelsen* (43): 1–77+map.

Lönnberg, E. 1899. 'Fortsatta undersökningar rörande Öresunds djurlif.' *Meddelanden från Kongliga Landtbruks-styrelsen* (49): 1–24.

Lönnberg, E. 1903. 'Undersökningar rörande Skeldervikens och angränsande Kattegat-områdets djurlif. Med ett tillägg om alg-vegetationen af Prof. F. R. Kjellman.' *Meddelanden från Kongliga Landtbruksstyrelsen* (80): 1–81.

Meisen, V. (ed.) 1932. *Prominent Danish Scientists through the ages*. Copenhagen: Levin & Munksgaard.

Meyer, H.A. & Möbius, K. 1865. *Fauna des Kieler Bucht. Erster Band: Die Hinterkiemer oder Opisthobranchia*. Leipzig: Wilhelm Engelmann.

Meyer, H.A. & Möbius, K. 1872. *Fauna des Kieler Bucht. Zweiter Band: Die Prosobranchia und Lamellibranchia nebst einem Supplement zu den Opisthobranchia*. Leipzig: Wilhelm Engelmann.

Molander, A.R. 1928. 'Investigations into the vertical distribution of the fauna of the bottom deposits in the Gullmar Fjord.' In: *Kristinebergs Zoologiska Station 1877–1927. Skriftserie utgiven av K. Svenska Vetenskapsakademien* 1930: 2. 1–90+map.

Molander, A.R. 1962. 'Studies of the fauna in the fjords of Bohuslän with reference to the distribution of different associations.' *Arkiv för Zoologi*, Ser. 2, 15(1): 1–64.

Moore, P.G. & Seed, R. (eds.) 1986. *The Ecology of Rocky Coasts*. New York: Columbia University Press.

Nejle, G. 1998. *Otto Torell. En skildring av varberssonen Otto Torells liv och gärning (1828–1900)*. Varberg: Gert Nejle & Hembygdsföreningen Gamla Varberg.

Nordenskiöld, A.E. 1882. *Vega-Expeditionens vetenskapliga iakttagelser*, vol. 1. Stockholm: F. & G. Beijers Förlag.

Nordenskiöld, E. 1935. *The History of Biology. A Survey*. New York: Tudor Publishing Co.

Nørrevang, A. & Lundø, J. (eds.) 1979. *Danmarks natur 3. Havet*. Copenhagen: Politikens Forlag.

Olsgard, F., Schaaning, M.T., Widdcombe, S., Kendall, M.A. & Austen, M.C. 2008. 'Effects of trawling on ecosystem functioning.' *Journal of Experimental Marine Biology and Ecology* 366: 123–33.

Ørsted, A.S. 1843. 'Annulatorum Danicorum Conspectus.' *Det Kongelige Danske Videnskabernes Selskab. Naturvidenskab og Mathematik. Afhandlinger*, 10: i–iv, 1–52, pls I–VII.

Pearson, T.H., Josefson, A.B. & Rosenberg, R. 1985. 'Petersen's benthic stations revisited. I. Is the Kattegatt becoming eutrophicated?' *Journal of Experimental Marine Biology and Ecology* 92: 157–206.

Petersen, A.H., Clausen, P., Gamfeldt, L., Hansen, J.L.S., Norling, P., Roth, E., Svedäng, H. & Tunón, H. 2018. 'The Sound: Biodiversity and ecosystem services in a densely populated and heavily exploited area.' In: *Biodiversity and ecosystem services in Nordic coastal ecosystems: an IPBES-like assessment. Volume 2. The geographical case studies*, ed. T. Tunón, TemaNord 2018:532. Copenhagen: Nordic Council of Ministers, 135–171.

Petersen, C.G.J. 1888. *Om de skalbærende molluskers udbredningsforhold i de danske have indenfor Skagen*. Copenhagen: Andr. Fred. Høst & Søn's Forlag.

Petersen, C.G.J. 1893. *Det videnskabelige udbytte af kanonbaaden 'Hauchs' togter i de danske have indenfor Skagen i aarene 1883–86*. Copenhagen: Andr. Fred. Høst & Søn's Forlag.

Petersen, C.G.J. 1913. 'Havets Bonitering. II. Om Havbundens Dyresamfund og om disses betydning for den marine Zoogeografi.' *Beretning til Landbrugsministeriet fra Den Danske Biologiske Station* 21:1–42.

Petersen, C.G.J. 1918. 'The sea-bottom and its production of fish food. A survey of the work done in connection with valuation of the Danish waters from 1883–1917.' *Beretning til Landbrugsministeriet fra Den Danske Biologiske Station* 25: 1–62.

Petersen, C.G.J. & Boysen-Jensen, P. 1911. 'Havets Bonitering. I. Havbundens Dyreliv, dets Næring og Mængde. (Kvantitative Studier).' *Beretning til Landbrugsministeriet fra Den Danske Biologiske Station* 20: 3–78.

Roller, D.W. 2014. *The Geography of Strabo. An English Translation, with Introduction and Notes.* Cambridge: Cambridge University Press.

Rosenberg, R., Gray, J.S., Josefson, A.B. & Pearson, T.H. 1987. 'Petersen's benthic stations revisited. II. Is the Oslofjord and eastern Skagerrak enriched?' *Journal of Experimental Marine Biology and Ecology* 105: 219–51.

Sandbeck, T. 2007. *Danske havforskningsskibe gennem 250 år.* Steenstrup: Skib Forlag.

Schouw, J.F. 1822. *Grundtræk til en almindelig Plantegeographie.* Copenhagen: Gyldendalske Boghandlings Forlag.

Spärck, R. 1933. *Dansk naturhistorisk Forening i København 1833–1933. Et dansk videnskabeligt selskab gennem hundrede aar.* Copenhagen: C.A. Reitzel.

Spärck, R. 1962. *Undervisningen i zoologi ved Københavns Universitet.* Copenhagen: Bianco Lunos Bogtrykkeri A/S.

Strand, J. & Jacobsen, J.A. 2002. 'Imposex in two sublittoral neogastropods from the Kattegat and Skagerrak: the common whelk *Buccinum undatum* and the red whelk *Neptunea antiqua.*' *Marine Ecology Progress Series* 244: 171–77.

Svane, I. & Gröndahl, F. 1988. 'Epibioses of Gullmarsfjorden: An underwater stereophotographical transect analysis in comparison with the investigations of Gislén in 1926–29.' *Ophelia* 28: 95–110.

Swedenborg, E. 1719. *Om Watnens Högd och Förra Werldens Starcka Ebb och Flod. Bewis Utur Svergie.* Stockholm: Johan H. Werner.

Söderqvist, T. 1986. *The Ecologists. From Merry Naturalists to Saviours of the Nation.* Stockholm: Almqvist & Wiksell.

Thorson, G. 1938. *Verdenshavenes dyreliv*. Copenhagen: Folkeuniversitetsudvalget.

Thorson, G. 1944. 'Hundrede aars Øresundsundersøgelser.' *Dyr i Natur og Museum 1943–44*: 41–59.

Uddenberg, N. 2004. *Idéer om livet. En biologihistoria*. Vols. I–II. Stockholm: Natur och kultur.

Waern, M. 1952. 'Rocky-shore algae in the Öregrund archipelago.' *Acta Phytogeographica Suecica* 30: 1–298.

Warming, E. 1881. 'Den danske botaniske literatur fra de aeldste tider til 1880.' *Botanisk Tidsskrift* 12: 42–247.

Warming, E. 1905. 'Ørsted, Anders Sandøe.' In: *Dansk biografisk lexicon tillige omfattende Norge for Tidsrummet 1537–1814*, vol 19, ed. C.F. Bricka, Copenhagen: F. Hegel and son, 390–93.

Wolff, T. 1967. *Danish Expeditions on the Seven Seas*. Copenhagen: Rhodos International Science and Art Publishers.

Wolff, T. 1979. *Københavns Universitet 1479–1979. Zoologi*. Copenhagen: G.E.C. Gads Forlag.

Worster, D. 1996. *De ekologiska idéernas historia*. Stockholm: SNS Förlag.

Table I

Several scientists have honored A.S. Ørsted by using his name in the scientific names of organisms (listed in alphabetical order). Only marine organisms are included.

Ageratum oerstedii B.L. Rob., 1913	Tracheophyta	
Allostoma oerstedi (Levinsen, 1879)	Platyhelminthes	
Amphiodia oerstedi (Lütken, 1856)	Echinodermata, Ophiuroidea	
Anamobaea orstedii Krøyer, 1856	Annelida, Polychaeta	
Aricia oerstedii Claparède, 1864	Annelida, Polychaeta	a synonym of *Protoaricia oerstedii* (Claparède, 1864)
Astropecten oerstedi Lütken, 1859	Echinodermata, Asteroidea	a synonym of *Astropecten armatus erinaceus* Gray, 1840
Batophora oerstedii J. Agardh, 1854	Chlorophyta	
Cephalothrix oerstedii Diesing, 1850	Nemertea	a synonym of *Cephalothrix rufifrons* (Johnston, 1837)

Cerebratulus oerstedii Beneden, 1861	Nemertea	a synonym of *Siphonenteron bilineatum* Meneghini in Renier, 1847
Chaetosyllis oerstedi Malmgren, 1867	Annelida, Polychaeta	nomen dubium
Chara inconstans var. oerstediana A. Braun, 1859	Charophyta	
Dasya trichoclados var. oerstedii J. Agardh [1863?]	Rhodophyta	uncertain
Dentalium oerstedii Mørch, 1861	Mollusca, Scaphopoda	
Ehlersia oerstedi (Malmgren, 1867)	Annelida, Polychaeta	nomen dubium
Elysia oerstedii Mørch, 1859	Mollusca, Gastropoda	
Epitonium oerstedianum Hertlein & A.M. Strong, 1951	Mollusca, Gastropoda	a synonym of *Epitonium replicatum* (G.B. Sowerby II, 1844)
Euclymene oerstedii (Claparède, 1863)	Polychaeta	
Eunice oerstedi Stimpson, 1854	Annelida, Polychaeta	
Eunice oerstedii Stimpson, 1853	Annelida, Polychaeta	
Eunoe oerstedi Malmgren, 1865	Annelida, Polychaeta	

Exogone oerstedii (Kölliker in Koch, 1846)	Annelida, Polychaeta	nomen dubium
Heteronereis oerstedii Quatrefages, 1866	Annelida, Polychaeta	a synonym of *Platynereis pulchella* Gravier, 1901
Hirtoscala oerstediana (Hertlein & A.M. Strong, 1951)	Mollusca, Gastropoda	a synonym of *Epitonium replicatum* (G.B. Sowerby II, 1844)
Leptuca oerstedi (Rathbun, 1904)	Crustacea, Decapoda	
Neogonodactylus oerstedii (Hansen, 1895)	Crustacea, Stomatopoda	
Nephthys oerstedii Quatrefages, 1866	Annelida, Polychaeta	a synonym of *Nephthys caeca* (Fabricius, 1780)
Oerstedia de Quatrefages, 1846	Nemertea	
Oerstediella Friedrich, 1935	Nemertea	a synonym of *Oerstedia*
Oerstediidae Chernyshev, 1993	Nemertea	
Oerstedtia Trevisan, 1848	Phaeophyceae	
Ophiactis oerstedi Lütken, 1856	Echinodermata, Ophiuroidea	a synonym of *Ophiactis simplex* (LeConte, 1851)
Pachychilus oerstedii Mörch, 1861	Mollusca, Gastropoda	
Panthalis oerstedi Kinberg, 1856	Annelida, Polychaeta	

Phyllodoce oerstedii Quatrefages, 1866	Annelida, Polychaeta	
Pisione oerstedii Grube, 1857	Annelida, Polychaeta	
Polyodontes oerstedi (Kinberg, 1855)	Annelida, Polychaeta	
Pomatoceros oerstedi Voss & Voss, 1955	Annelida, Polychaeta	a synonym of *Spirobranchus giganteus* (Pallas, 1766)
Pseudogeoplana oerstedi (Graff, 1899) Ogren & Kawakatsu, 1990	Platyhelminthes	
Selene oerstedi Lütken, 1880	Chordata, Teleostei	a synonym of *Selene orstedii* Lütken, 1880
Sipunculus (Phascolosomum) oerstedii Quatrefages, 1865	Sipunculoidea	a synonym of *Golfingia (Golfingia) margaritacea* (Sars, 1851)
Solen oerstedii Mörch, 1860	Mollusca, Bivalvia	
Syllis oerstedi (Malmgren, 1867)	Annelida, Polychaeta	nomen dubium
Tanais oerstedii Krøyer, 1842	Crustacea, Tanaidaceae	a synonym of *Heterotanais oerstedii* (Krøyer, 1842)
Tetragonopterus oerstedii Lütken, 1875	Chordata, Teleostei	a synonym of *Psalidodon fasciatus* (Cuvier, 1819)
Tetrastemma oerstedi Senz, 2001	Nemertea	

DE REGIONIBUS MARINIS.

ELEMENTA

TOPOGRAPHIÆ HISTORICONATURALIS FRETI ÖRESUND.

DISSERTATIO INAUGURALIS,

QUAM

AD HONORES MAGISTRI ARTIUM RITE CAPESSENDOS

SCRIPSIT

ET

DIE XXIX MENSIS APRILIS.

RESPONDNETE ORNATISSIMO

E. PETIT.

CHIR. MIL.

PUBLICE DEFENDERE STUDEBIT

A. S. ØRSTED.

PHIL. CAND.

HAUNIÆ.

Typis excussit J. C. Scharling.

MDCCCXLIV.

1844

Dissertationem hanc Facultas philosophica in Universitate hauniensi dignam censet, quæ pro gradu magistri rite obtinendi publico Eruditorum examini subjiciatur.

I. Reinhardt,
h. a. Decanus Facult. philos.

PATRUIS OPTIMIS DILECTISSIMIS

A. S. ÖRSTED,

QUI MIHI A PRIMA PUERITIA LOCO PATRIS BENEVOLENTISSIMI FUIT

ET

H. C. ÖRSTED,

QUI MEORUM STUDIORUM OPTIMUS DUX SEMPER FUIT

QUOS EXEMPLA PRÆSTANTISSIMA OMNIS BONI, EXCELSI, EXIMII SEMPER SUSPEXI,

HANC DISSERTATIONEM

SACRAM ESSE VOLUIT

Auctor.

Præmonenda.

Sæpe mihi, secundum littora freti Oresundici commigranti, et varietatem characteris terræ et vegetationis libenter adspectanti, quum altera ex parte æquor uniforme oculis obversans varietatem, quam in fundo maris esse sciebam, semper occultaret, maxime optandum videbatur, ut ad caliginem amovendam, qua tanta superficiei terræ pars erat obducta, et tot eximia membra organismorum naturæ tecta, aliquid afferrem. Quamquam fretum mihi animalia plantasque multas et novas et raras iam præbuerat, hoc tamen parum contentus eram; deerat enim universalis imago, quæ tum demum existit, quum singula ita connectuntur, ut totam rem amplectamur, et in varietate unitas quoque appareat. Organismos numerosos per fundum maris non temere dispersos, sed certis legibus distributos esse, opinabar. Quum igitur totam æstatem a. 1842 *investigationibus illis, quarum studio animus iamdiu ardebat, impendere mihi licuisset, spem meam non fefellerunt fructus, quos ex methodica maris pervestigatione percepi. Quæ igitur ipse fretum Øresundicum pervestigando, ad summam rei pertinentia, intellexi, ea cum aliis communicare ausus sum hac commentatione, quæ elementa topographiæ historiconaturalis freti continet. „De regionibus marinis" inscriptus est hic libellus, quia universa ratio geographica animalium et plantarum in freto ut regiones apparet, cum fretum tam angustum sit, ut fere ex duobus solummodo lateribus declivibus constet. Vox enim regionis hic ita ut hodie a plerisque auctoribus usurpatur. Extensio igitur est directionis verticalis. Ubi de ratione geographica animalium agitur, hæc vox idem significat. Geographiam enim plantarum priorem in systema redactam esse, ita geographiæ animalium utile et fructuosum est, ut non modo totius fere systematis adumbrationem, sed etiam multos terminos, qui hodie ab omnibus adoptati sunt, inde repetere possimus. Itaque geographia plantarum geographiam animalium miro modo quasi præparavit.*

Hunc libellum in duas partes dividere, quarum altera generalia, altera enumerationem quam plenissimam omnium plantarum animaliumque, quæ in freto reperta sunt, cum indicatione habitationum et descriptione atque iconographia novarum specierum, quæ non paucæ sunt, contineret, primum constitueram. Veritus autem, ne pars descriptiva maior evaderet, quam quæ dissertationi conveniret, optimum putavi, enumerationem non copiosiorem addere, quam necesse esset, ut organismi, qui in freto invenirentur, cum aliis comparari possent, et ad species solummodo novas, quæ imprimis descriptione dignæ essent, breves diagnoses afferre. Fieri potest, ut alio loco, quæ desunt, expleamus.

Mihi gratissimum est, h. l. vv. cll. Hofmann Bang, I. Agardh, Liebmann, et Drr. Beck et Kröyer publice gratias agere, quia tres priores de algis, duo posteriores de animalibus me in multis speciebus diligentius constituendis adiuverunt.

Gaudium summum, quod ex freto pervestigando percepi, quum singula quasi coalescere et ad unum tendere inciperent, unitasque varietatum clarius appareret, ii soli, qui in natura perscrutanda versantur, intelligere possunt. Itaque tempus, quo fretum visitavi, gratissimum vitæ meæ semper recordabor, et vehementissime cupio, occasionem mihi præberi, ut cetera quoque maria, insulas Danicas cingentia, investigem.

De enumeratione hæc notanda:

1. *Si nulla speciei alicuius habitatio indicata est, significatur, eam maiori parti regionis, ad quam relata sit, communem esse.*
2. *Quum duo loca lineola interiecta (velut Kullen-Landskrona) nominantur, habitatio specierum ultima septentrionem et meridiem versus significatur.*
3. *Ubi nomen eius, qui speciem aliquam reperit, omissum est, ipse loco indicato eam reperi.*
4. *Paucas species, quas ex locis extra fretum sitis huc transtuli, fretum quoque sine dubio habet.*

Argumentum libelli.

Caput IV.

INTRODUCTIO HISTORICA[1])

Non eadem celeritate omnes artes excoli possunt. Levis cognitio historiæ artium ostendet, quomodo numerus illarum paullatim auctus sit, et quomodo altera semper alteram præparaverit eique viam muniverit. Hoc vero aliter fieri non potuit. Argumento est ars quæ nobis leges rationis, quam plantæ et animalia cum superficie terræ habent, sive doctrinam, quomodo statio, extensio distributioque eorum ad momenta externa, quibus afficiantur, respondeant, i. e. Geographiam plantarum et animalium tradit.[2]) Antequam enim et Climatologia et Physiologia animalium et plantarum ad quendam perfectionis gradum venit, et antequam investigationibus faunarum et florarum plenior cognitio distributionis organismorum per totum orbem terrarum parta est, illæ artes omnino existere non potuerunt. Sic extremo seculo priore geographia plantarum itinerariis et floris (inprimis Tournefortii voyage au Levant et Linnéi flora lapponica) quasi præsagiri coepta est.

1) Etsi in hoc libello tantum parvula geographiæ animalium pars pertractatur, non alienum esse putavi, conspectum historicum hujus scientiæ evolutionis et status præsentis præmittere.

2) Schouw Grundtræk til en almindelig Plantegeographie p. 5.

A. MDCCCVII demum in Humboldtii „Essai sur la géographie des plantes“ et in „Tableau des regions equinoctiales“ propria ars orta est. Secuta sunt tot præstantissima huius artis auxilia, maxime Wahlenbergii flora lapponica, Roberti Brownii „general remarks on the botany of terra australis“ a. MDCCCXIV, Humboldtii dissertatio „sur les lignes isothermes“ a. MDCCCXVII, Decandollii in Memoires de la societé d'Arcueil, ut Schouw nostras Professor doctissimus a. MDCCCXXII certum huius artis systema[1]) condere posset. Dum hoc modo geographiæ plantarum propria ars iamdiu est, de geographia animalium hoc ne hodie quidem dici potest; quamquam enim multa et egregia auxilia collecta sunt, nondum tamen sunt in systema redacta. Cuius rei causa in magna inopia operum faunicorum inesse videtur, quæ tamen inopia, quod ad animalia superiora, sed neutiquam ad inferiora pertinet, recentioribus temporibus sublevata est.

Zimmermannus[2]) primus gravia auxilia distributionis animalium geographicæ attulit, et (in tertio volumine) ita in universum hanc rem adumbravit, ut illo tempore nihil rectius merito exspectari posset. Apud Treviranum[3]) maior copia in multis rebus apparet, sed investigationes, quæ antecedebant, pauciores erant, quam quibus niti posset disciplina, quæ posteritati quoque satisfaceret. Illiger iam a MDCCCXI (in Act. Academ. Berolin) optima plenissimaque de distributione mammalium aviumque tradidit. Quod Anthropologiæ Rudolphi (1812) caput „Ueber die Verbreitung der Thiere“ inscribitur causas solas indicat,

---

1) Schouw op. cit.

2) Geographische Geschichte des Menschen und der allgemein verbreiteten vierfüßigen Thieren mit einer zoologischen Weltkarte Vol. 1—3. Leipz. 1778—83.

3) Biologie 2 Bd. Göttingen 1803.

quibus animalia primitus uno loco congregata fuisse, unde in diversissimas orbis terrarum regiones dispersa fuisse, non posse statui, probatur. *Latreillii* commentationes de distributione geographica insectorum[1]) magni momenti sunt ad leges distributionis animalium universæ illustrandas. *Fabricius* iam antea terram in octo regiones zoologicas secundum faunas diversas insectorum dividere conatus erat; *Lesson* opusculum Illigeri de avibus et *Minding* eiusdem scripta de distributione mammalium[2]) supplere nisi sunt. *Swainson* copiosissima de terra in regna zoologica dividenda adhuc tradidit.[3]) Terram secundum aves in sex regna: europæicum, asiaticum, americanum, africanum, australicum, arcticum, dividit. Liber *Prichardii* de homine[4]) optimam adhuc descriptionem distributionis regni animalium geographicæ in prolegomenis continet. Distributio geographica singularum animalium classium pertractatur, terraque maxime secundum mammalia et reptilia, quæ quidem optimum fundamentum talium divisionum iaciant, in has novem provincias zoologicas dividitur: regnum arcticum, Americam temperatam, Europam Asiamque temperatam, Africam tropicam et australem, Asiam tropicam, Archipelagus Indicum, Polynesiam, Australiam. Distributionis animalium terrestrium hæ generales leges evolvuntur: 1) zona arctica sola, in qua tres partes orbis terrarum unum regnum complectuntur, easdem species habet; 2) in ceteris regnis, quæ maribus segregantur, nunquam eædem species in-

---

1) Introduction à la géographie général des Insectes et des Arachnides. Memoire du Museum d'histoire naturelle 1815.

2) Geographische Verbreitung der Säugthiere 1829.

3) A Treatise on the Geographic and Classification of Animals. London 1835.

4) Researches into the physical history of mankind 1836.

veniuntur, interdum eadem genera, sæpissime analoga genera et familiæ; 3) diversæ species in hemisphæris duabus; 4) parvæ insulæ, procul a continente sitæ, nulla indigena mammalia habent; 5) insulæ prope continentem sitæ eadem genera animalium, quæ continens, habent. Unde concluditur, animalia in regionibus, ad suam vivendi rationem aptis, indigena esse, nec ab uno loco, ut Linné putavit, profecta esse. *Schlegel*[1]) distributionis serpentium geographicæ descriptionem dedit et chartam summi momenti, non solum quod materiam pæne exhaurit, sed etiam quod distributio horum animalium per terram magis, quam ceterorum, ad quæstionem de domicilio animalium primario illustrandam apta est, quum illa difficilius terram, quam habitant, relinquere possint, nec causa, cur homines eas ex illa deportent, appareat, nec facile exstinguantur. Secundo volumine *Lacordairii* Introduction à l'Entomologie Paris 1835, ratio insectorum geographica copiose describitur. *Lyell* in Principes of Geology Vol. III the fifth Edition 1837, multa notabilia ad extensionem animalium illustrandam affert. *Oswald Heer* ad illustrandum, quatenus color animalium climate afficiatur, complura gravioris momenti contulit. In *Berghausii* „Allgemeine Länder- und Völkerkunde 3 Band 1838" copiosa, sed „informis" compilatio geographiæ animalium inest. *R. Wagner* in versione Prichardii „physical history of mankind" compendium librorum Swainsonii et Schlegelii, quos supra commemoravi, addidit.

In distributione autem geographica animalium *marinorum* constituenda multa quidem desunt. Iam *Zimmermannus* recte dixit, in mari, quod temperatura minus variaret, inter diversos latitudinis gradus non eandem ac

---

1) Essai sur la physionomie des serpens. Amsterdam 1837 Bd. 1 p. 195.

in terra esse animalium diversitatem, sed audacius iusto contendit „ein Seethier ist oder könne doch in Meer fast alle Orten zu Hause seyn" (tom. 3, p. 218). Quæ sententia falsa ex eo, quod species analogæ ad polum utrumque identicæ putabantur, maxime orta est. Hanc falsam esse, *Peron et Lesueur* primi animadverterunt, viginti diversas species sub nomine Phocæ ursinæ descriptas esse, eandemque perturbationem in specie Phocæ vitulinæ reperiri, ostendentes. Iidem Stellerum et Fabricium diversissimas species sub nomine Phocæ leoninæ descripsisse probarunt, et meliora nos docuerunt de distributione animalium marinorum, contendentes, in maribus arcticis et antarcticis analogas quidem, non vero identicas species reperiri („de tous les animaux que nous avons pu voir nous-même, il n'en est pas un seul qui ne se distingue essentiellement des espéces boreales analogues". Annales du Museum d'hist. nat. Tom. 15 p. 300 1810). Iidem autem sententiam falsissimam de statione corallorum perferre conati sunt, opinionem veterem, multas insulas maris australis ex corallis ortas esse (Voyage II, pag. 165—192) affirmantes atque augentes. *Quoy et Gaimard* bene meriti sunt, quod primi hanc sententiam refutantes, coralla, crusta exiguæ crassitudinis rupes tegentia, infra altitudinem paucarum orgyiarum non inveniri, ostenderunt, nec, verisimile esse, perspicue probarunt, eadem animalia vivere posse in ea pressus marini diversitate, in qua, si insulas a profunditate 1000—1200 pedum ad superficiem construxissent, victura fuerint. (Annales des sciences naturelles 1825). Postea *Ehrenbergius* in vestigationibus suis in mari rubro veritatem observationum illarum confirmavit, ubique reperiens, tenuem crustam crassitudine non plus 1—4 pedum corallis formari, eademque altius 18 pedes raro inveniri, ita ut nequaquam insulas extruere, sed sæpe quidem ab effectu fluctuum pernicioso eas defendere possent[1]. *Milne Edwards*, primus diversitatem animalium in diversis maris profunditatibus certius definire conatus, quatuor regiones secrevit[2]. Idem intellexit *Sars* investigationibus littorum Norvegiæ[3]. Ehrenbergius distributionem animalium infusoriorum geographicam copiosissime illustravit[4]. Plenissima et copiosissima omnium, quæ de maritima ullius animalium classis distributione scripta sunt, continet Milne Edwardii de Crustaceis liber[5]. Mare secundum Crustacea in tredecim regna dividens, hæc resultata generalia ex pervestigationibus assequitur: 1) numerus specierum a polis ad æquatorem augetur; 2) formæ perfectissimæ in maribus tropicis inveniuntur; 3) multitudo individuorum in maribus arcticis summa est; 4) creatio specierum ex diversis centris exisse videntur, nec ex uno loco, quod Linneus putavit; 5) magna maria secernentia extensionem specierum circumscribere videntur. Nullæ igitur species littoribus tropicis maris Atlantici utrinque communes sunt. 6) Oræ longæ, series insularum magnæ, flumen extensioni specierum favent. Recentioribus temporibus apud nos dissensio de extensione geographica cetaceorum exstitit, quum Eschricht sententiam veterem defenden-

---

1) Act. Acad. Berol. 1832. Ueber die Natur und Bildung der Korallenbänke des rothen Meeres.

2) Annales des scienc. natur. Tom. 21 p. 326.

3) Beskrivelser og Jagttagelser over nogle mærkelige Dyr o. s. v. Bergen 1835 p. VI.

4) Geographische Verbreitung der Infusionsthiere. Monatsbericht der Berliner-Academie 1840 p. 157—197.

5) Annal. des sciences natur. Tom. X et in histoire naturelle des Crustacés Vol. III, pag. 555.

dam, complura cetacea cosmopolitas esse, susceperit[1]) quam sententiam Kroyer refutare contendit[2]).

---

[1]) Förhandlingar vid de Skandinaviske Naturforskarnes tredje Möte i Stokholm 1842 p. 203.

[2]) Kroyers Naturhistoriske Tidsskrift 5 B. 5 H. p. 474.

CAPUT PRIMUM.

Rationes generales physico-geographicæ freti Øresundici.

Fig. 1.

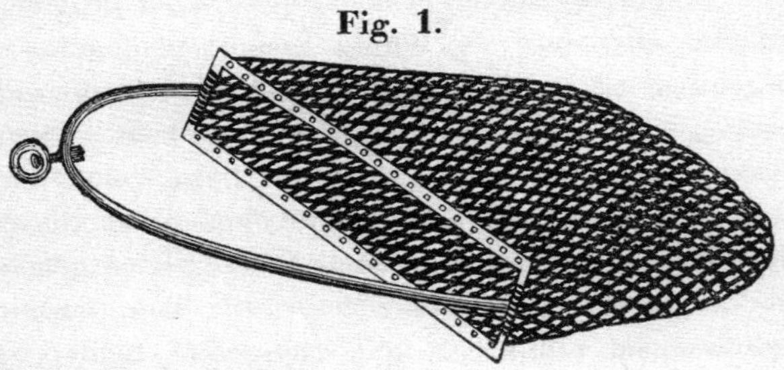

De methodo investigationis.

§ 1.

Quum fretum Öresundicum et animalium plantarumque, quæ in eo essent, distributionem geographicam accuratissime cognoscere studerem, nec aliæ fundi partes, quam quæ littus attingunt, observationibus meis ita faverent, ut mea ipsius ope contentus essem, ad maximam freti partem pervestigandam peculiaribus instrumentis peculiarique methodo uti necesse fuit. Ad talia investiganda quam fuit utile naturam Piscecolæ, natatoris insignis, induere et in fundo maris quasi in terra circumvagari, aut saltem campana urinatoria demergi posse! Quibus deficientibus, radula Ballii (den Ballske Skrabe) ad res ex fundo tollendas, quæ oculis conspici non possunt, aptissima est. Adumbratio eius hic addita est, ex qua intelligitur, eam pæne eodem modo, quo instrumentum ad ostreas captandas usitatum[1]) constructam esse. Constat ex forma ferrea rectangulari, quæ ex adverso instructa est ferreo arcu, ad duo latera minora affixo, et a tergo reti, cuius fila crassa commissuris angustis nexa sunt. Medius arcus annulum habet, qui axiculo circumagitur. Annulo funis affigitur. Radula demersa a navicula leniter progredienti pendens, fundum quasi corradit. Ita fit, ut parte fundi soluti, et plantis animalibusque, quæ in eo inveniuntur, paullatim impleatur. Si fundus maiore profunditate investigandus est, pondera plumbea, gravitatem radulæ augentia, imis arcus brachiis adnectenda sunt. Semper adverso flumine trahenda est, ne quid eam compleri prohibeat. Radula trîangularis aut rotunda eadem constructione utilissima est, si in altiores fundi partes penetrare velis. Neque vero semper facile est hanc machinam simplicissimam adhibere, quia sæpe aliud fundi, aliud summi maris flumen est, quia exercitatione quadam opus est, ut animadvertas, satisne an nimis fundum machina premat, quod sæpissime ex longitudine funis pendet. Qua in re te non poenitebit piscatores consuluisse, retia in fundo versare assuetos.

Quum intellexissem, fieri non posse, radula temere sine ordine et ratione demersa, ut veræ summamque rei amplectentes leges organismorum in fundo distributionis invenirentur, certum consilium certamque freti perscrutandi rationem inii.

Lineis igitur transversis ab ora Sjællandica ad Scanicam a boreali limite freti ad australem constitutis, naturam fundi quam plurimis locis investigavi. Ubique utrum fundus argillosus an arenosus esset, quæque ibi reperirentur animalia et plantæ notavi. Situs uniuscuiusque loci,

---

[1]) v. H. Krøyer: De danske Østersbanker p. 74 Tab. I.

ubi peculiare aliquid apparebat, pyxi nautica signisque piscatorum in littore, quæ ubique in freto facile observari possunt, diligentissime designatus est, qua in re piscatores magnopere exercitati operam optimam præbent. Hac multorum locorum investigatione nititur tabula geographica, quam postea adumbrare conabor (Tab. I.)

De finibus, divisione profundidate freti.

§ 2.

Fretum Öresundicum maximum fretorum est, quibus cum mari baltico sinus Codanus coniungitur. Inter 55° 20′ lat. bor. et 56° 18′ lat. bor. situm a meredie septentrionem versus patet. Septentrionalis terminus est linea a Gilleleie ad Kullaberg, occidentalis ora Siællandica a Gilleleie ad Stevns, orientalis ora Scanica a Kullaberg ad Falsterbo, miridiem versus linea a Stevns ad Falsterbo ducta. Longitudo summa c. tredecim milliariûm, latitudo summa inter Havniam et Malmö c. quatuor, minima inter Kronborg et Helsingborg 6340 cubitorum. Ab ipsa natura in tres partes dividitur: septentrionalem, mediam, australem. *Pars septentrionalis*, ab imo sinu Codano ad Helsingoram et Helsingburgiam patens, quæ eadem pæne indole est ac sinus Codanus, Codana quoque pars freti Sundici vocari potest. *Media pars* linea, septentrionem versus inter Helsingoram et Helsingburgiam, meridiem versus inter Havniam et Barsebæk ducta terminatur. Tot a partibus inclusa maxime cum sinu natura congruit. *Pars australis* a termino australi mediæ partis ad imum fretum patens, cum mari Baltico omnino congruit, ideoque pars Baltica vocari potest. Media igitur pars proprium fretum est, quum duæ reliquæ maria, cum freto coniuncta, quasi præparent.

Profunditas universa freti tam parva est, ut in recta proportione ad latitudinem, sectionibus transversalibus cogitatis, nisi maxima mensura constitutis, exponi non possit. In sectionibus tabulis I & II exhibitis, in quibus uni milliario duo digiti respondent, profunditatis proportio ad latitudinem non recte apparere potest; ita enim septem orgyiæ essent modo 571ma pars duorum digitorum, quæ quidem magnitudo conspici non posset. Hic vero septem orgyiæ uni lineæ æquales sunt, ut sola ratio relativa profunditatis diversorum freti locorum exhibeatur; sed profunditas horum omnium, cum latitudine comparata, nimia est. Ex his sectionibus intelligi potest, modo in australi freti parte (Tab. II, 6) medium maiorem habere profunditatem, eandem vero et in boreali parte (Tab. I, 1) et in media (Tab. I, 2—4, Tab. II, 5) maxime Sueciam versus esse; item profundidatem latitudini retrorsum respondere; ita in latissima freti parte eam in medio non amplius orgyiarum octo esse (Tab. II, 6); in boreali summam orgyiarum quindecim (Tab. I, 1), in media angustissimaque parte vero inter Hveen et Scaniam maximam i. e. orgyiarum viginti quinque esse (Tab. I, 4)[1]).

De compositione chemica et pondere specificio aquæ.

§ 3.

Partes diversæ freti Öresundici diversitatem summam ponderis aquæ specifici, vel quod idem est, multitudinis

---

[1]) conf. Speciel-Kort over Sundet 1840 oprindelig efter Admiral Klint.

salis, qui in ea solutus est, ostendunt. Quæ diversitas inter partem australem et septentrionalem eadem esse videtur quæ inter mare Balticum, et septentrionale est, ideoque credo observationibus ponderis specifici utriusque maris investigationes ipsius freti facile expleri posse. Marcet hæc resultata exponit:

Mare Balticum

Pondus specif.	*Altitudo.*
1,0061	superficies.
1,0171	septendecim orgyiæ.
1,0272	quatuordecim orgyiæ.

Idem auctor has rationes inter septentrionalem partem maris Atlantici et Baltici affert: aqua illius 1,02886 pond. specif., et 4,26 p. C. salis habet, huius vero 1,00490 et modo 0,66 p. C. salis[1]). Diversitas igitur ponderis specifici horum marium est = 0,02396, et salis est = 3,60; eandemque inter partem australem et borealem freti esse, merito statuere possumus. Etiam eodem freti loco diversitas salsitudinis aquæ a directione fluminis ita pendet, ut facillime etiam gustu observari possit. Nec sine fructu atque utilitate erit, si quis diversitatem ponderis aquæ specifici, et flumine boreali et flumine australi (velut ad Havniam) accurate examinaverit.

De temperie aquæ freti.

§ 4.

Observationes temperiei aquæ freti proximis solis annis in munimento Trekroner institutæ in conspectu

1) Gehlers physicalisches Wörterbuch Artkl. Meer 6 Bd.

Act. Acad. reg. Havn. scient. expositæ sunt. In hoc vero temperies media sola diurna indicata est; quare ex his observationibus temperiem mediam et totius anni 1842 et singulorum mensium extraximus, ut eam cum temperie media aëris ejusdem anni comparare possimus.

	Temp. aëris.		Temp. aquæ.
Jan.	1,°4	....	1,°8
Febr.	0,2	....	1,2
Mart.	2,5	....	2,3
April.	5,4	....	4,2
Mai.	10,2	....	9,5
Jun.	12,2	....	11,5
Jul.	12,9	....	13,4
Aug.	15,7	....	16,5
Sept.	11,2	....	13,0
Oct.	6,3	....	8,0
Nov.	1,4	....	3,8
Dec.	3,0	....	3,8
Tot. an.	6,8	....	7,4

Ex his intelligitur temperiem mediam totius anni aquæ 0,6 majorem esse quam aëris, et hujus rei causam esse majorem aquæ temperiem per hiemem; nam mensibus Mart. — Iul. incl. aer aquam temperie superat.

De Flumine.

§ 5.

Flumina, quæ in freto reperiuntur, localia sunt, i. e. cum fluminibus maris universi directe non cohærent. Natura fluminis eadem est, quam plerumque habent freta angusta, quæ mare inclusum, in quod magna multitudo aquæ dulcis confluit, cum maioribus maribus, quibus illud circumdatur, coniungunt. Nisi igitur alia momenta eodem tempore directionem fluminis afficerent, semper flumen lene aquam e mari Baltico in Codanum duceret, ut æquor utriusque maris in eadem altitudine esset; itaque hic ex directione freti boreali-meridionali flumen semper australe esset. Æquor enim Baltici maris multo altius æquore Germanici est; ex auctoritate Nordenankeri[1]) 8′ altius, auctoritate Woltmanni ad Kiliam 1′ 2″[2]). Observationibus etiam in navali regio factis compertum est, flumen australe boreali bis et dimidio frequentius esse[3]). Ventus igitur propriæ fluminis directioni, australi, renititur. Soli tamen aquilo et corus, quorum prior nostra in terra rarior est, flumen boreale, ceteri venti australe efficiunt. Directioni igitur fluminis australi venti quoque favent[4]). Secundum observationes Klintii, præfecti classis, pondus aëris, quod barometro indicatur, nec a directione venti directe dependet, diversitatem altitudinis æquoris (ad 5′) notabilem afferre, ergo etiam directionem celeritatemque

---

1) Von den Strömen der Ostsee. Leipz. 1795. 8.

2) Poggendorffs Annalen II, 444.

3) Schouw, Skildring af Veirligets Tilstand i Danmark 1826; p. 526 dicitur, diversis freti locis flumen observari necesse esse, quum animadvertendum sit, alium flumen in medio freto, alium ad latera esse; sed etiam necesse erit, directio fluminis in altitudine diversa examinetur; vulgo enim compertum est, et pæne quotidie quidem a piscatoribus, qui retia in fundum demergunt, ad hunc plerumque in maiore profunditate flumen australe esse, quamquam in superficie boreale sit. Fortasse igitur ut lex generalis ponendum est, *in freto semper esse flumen australe*; quum enim venti boreale efficiunt, modo in superficie adest.

4) Den danske Lods p. 104. 1843.

fluminis afficere potest. Æquoris altitudo vulgaris maxima est mense Sextili, minima Aprili. Cuius rei causa apparet; ipsis enim mensibus, quibus aqua altissima est, flumen australe minus frequens quam boreale reperitur, et quum borealis prohibet, ne abundantia aquæ maris Baltici effluat, æquor et in hoc et in freto altius sit, necesse est.

Commutationes periodicæ altitudinis æquoris, quæ motu lunæ circum terram quotidiano efficiuntur, in septentrionali freti parte, quæ sinum Codanum attingit, paullulum, in media infimaque omnino non animadverti possunt.

CAPUT SECUNDUM.

Rationes freti Øresundici geologicæ.

§ 6.

Quamquam fundi natura geologica hic præcipue contemplanda est, necesse tamen erit, oræ quoque ratio geologica examinetur, partim quia fundus directa eius continuatio habenda est, partim ut quæstiones, quo tempore fretum formatum sit, quasque mutationes post formationem subierit, expediantur. Quas quæstiones nunc instituturi sumus.

Quo tempore et quomodo fretum formatum est?

§ 7.

Fretum quum strata periodi tertiariæ persecet[1]) ante extremam hanc periodum formari non potuit. Tempore antehistorico nostræ periodi, vetustissimis alluviis marinis (Havstokke) Bornholmii, et diluvie Cimbrica nondum factis, illud formatum esse, cl. Forchhammerus putat, sagacissimam coniecturam ad ortum freti Öresundici, aliorum fretorum sinuumque Daniæ explicandum faciens. Censet enim, sinum Botnicum, terris quæ circumiacent

---

1) v. geognostisk Kort over Danmark af Forchhammer 1843.

allevatis, in lacum commutatum esse, quo lacu viam abundantiæ aquarum, quæ ex regionibus circumiacentibus defluerent, per formationes minus solidas, quibus tunc Sueciæ pars australis cum Russia coniungeretur, patefactam esse. Geognostes ille præclarissimus iam putat, quum vis aquæ violenter exagitatæ pluribus locis perrumpendo viam sibi muniret, etiam fretum tali perruptione formatum esse [1]). Ut hoc modo formatio freti explicetur, certe videtur necesse esse statuere, id maxima ex parte antea fuisse, ex sinibus vero constans, quorum alter ad septentrionem, alter ad meridiem vergeret, et Siællandiam Scaniamque modo una vel forte duabus lingulis angustis coniunctas fuisse. Si configuratio harum terrarum, natura orarum ex adverso sitarum, distributio insularum vadorumque freti respiciuntur, verisimillimum mihi videtur, illas oras et ad Helsingoram et Helsingburgiam, et ad Havniam et Malmö, coniuntas fuisse, et inter has lingulas lacum patuisse. Si totum spatium freti terra fuisse putatur, non facile intelligitur, quomodo perruptio una diluvie fieri potuerit.

Quas mutationes fretum post formationem subiit?

§ 8.

Ad mutationes freti constituendas natura orarum pervestiganda est; quare in sequentibus brevem eius conspectum adumbrare conabimur. Etiam hac in re ad tres freti partes summa diversitas orarum apparet; ab australi enim

---

1) Skandinaviens geognostiske Forhold, et Foredrag holdt d. 22 Novbr. 1843 af G. Forchhammer.

ora Siællandiæ, quæ fretum attingit, incipientes, a septentrionali parte Stevnii (a loco, qui dicitur Bøgeskov) c. ad diversorium, quod Flaskekro vocatur, multas inveniemus series alluviorum marinorum[1]) quæ, cum ora parallela, octavam vel quartam partem milliarii in terram patentia agros steriles, qui „Heden" vel „Lyngen" vocantur, præbent. In parte oræ aquilonem versus alluvia ex lapide, in ceteris partibus ex arena et zostera marina constant, quæ diversitas ex eo orta videtur, quod has ora Suecica melius contra impetum undarum tegit. In maiore parte huius spatii alluvium invenitur, quod magnitudine insigne præ ceteris eminet tamquam agger, cuius australis pars, modo ex lapide constans, tam artificiosa apparet, ut, qui circum habitant, eum arte instructum esse putent; sed perspicuum est, eum violentissima tempestate formatum esse. Quum igitur ad australem freti partem ora hac terræ periodo magnopere aucta sit, fretum modo alluvione, non allevatione aliqua contractum videtur.

Pars oræ Siælandicæ mediam freti partem cingens perspicue ostendit, se mari allui, quod multo lenius agitatur, quam mare oræ australis et borealis; alluvia enim, hic reperta, perparva sunt. Præterea alluviones et abluviones secundum variam oræ naturam variantur; ubi hæc plana est, illæ, ubi prærupta, hæ inveniuntur. Abluviones ad Vedbæk inter Sletten et Humlebæk et ad Snedkersteen factæ sunt, quod ex eo intelligi potest, quia piscatores truncos arborum cum radicibus arenæ affixis in mari procul ab ora, quæ nunc est, repererunt.

Vis maris vehementius exagitati inprimis ad totam oram borealem comparet, quæ, exceptis paucis eminentibus locis, ad Hellebæk et Nakkehoved, ubi ora prærupta

---

1) Havstokke, Geschiebebänke.

est, alluvio, maxime ex arena constante, aucta est. Modo ad Hornbæk formatio alluvionum per multas series parallelas, quæ ex arena lapidibusque constantes seriebus littoris australis similes sunt, obvenit. Vis autem arenæ maxime ad Villingebæk conspicua et notabilis est, quia phænomenis elevationis simillima est. In colle enim, ab ora remoto, per quem via ducta est, stratum fossilium, (fig. 2 b) quæ specierum nunc mare habitantium sunt, Buccini undati, Fusi antiqui, Littorinæ littoreæ, Cardii edulis, Cyprinæ Islandicæ et aliorum, magna multitudine stratorum arenæ horizontalium (fig 2 aa) inclusum cernitur, ut primo aspectu effectus maris invenire nobis videamur. Diversa vero animalia, quæ in tam diversa profunditate vivunt, quam Cardium edule, Cyprina Islandica, res artifisiosæ, quæ ibi adsunt, inprimis magna carbonis multitudo, aliaque his similia, facile nobis persuadent, hunc collem cumulum arenæ, qui Klit appellatur, esse, ut ostendat, quantopere effectus aëris maris effectus imitari possint. Simile quid secundum ripam amnis qui, ad Villingebæk effluit, observari potest; in fundo enim amnis usque ad octavam milliarii partem ab ostio ejus magna multitudo testarum molluscorum marinorum, velut Cardii edulis, Tellinæ calcareæ, Nassæ reticulatæ, Littorinæ littoreæ aliorumque facile animadvertitur. Hæc res, quæ tam singularis videtur, fodiendis agris propinquis expeditur; tum enim sub strato arenæ, unum pedem alto, stratum horum animalium marinorum, interdum dimidium pedem altum, invenitur, quod

Fig. 2.

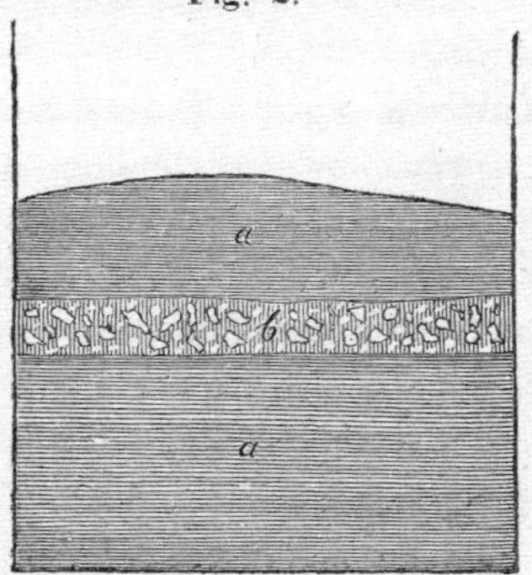

amnis perrumpens testas nudat. Facile perspicitur, quibus locis ab amne perrupti agri hodie sunt, quondam sinum maris fuisse, qui arena fere eodem modo, quo Hanveile et Bygholmveile, duo recessus sinus Liimfjorden, oppletus sit[1]). Lyngbye, Sacr. Min., animalia marina sub iisdem condicionibus ad Söborg reperisse dicitur; unde verisimile fit etiam hic sinum maris usque ad Söborg penetrasse. Quin lacus, qui „Arresö" vocatur, sinus maris fuerit, dubitari non potest; hic quidem vis arenæ violentior fuit.

Intelligimus igitur, mutationes, quas ora Siællandica subierit, etiam hodie continuari, quum ex alluvionibus et abluvionibus constent. Si forte mutationes violentiores inciderunt, necesse est, allevatio parvula fuerit; utrum enim magna altitudo super mare alluviorum marinorum tempestatibus vehementibus effecta sit, an simul allevatio terræ id effecerit, decernere non audeo.

Pars oræ Scanicæ, quæ fretum cingit, multo maius spatium abluvione quam alluvione mutatum esse, perspicue indicat. Hoc de ora, a Falsterbo ad Landskronam patente, quam quidem ipse non perscrutatus sum, valere, vulgo narrant[2]), ceteram autem sic ablui, ipse comperi. Hoc maxime accidit præruptæ illi oræ, quæ a Landskrona septentrionem versus extra Glumslöf longissime patet, ubi quondam prope mare lacus fuisse videtur; in summo enim prærupto et complura strata calcis aquæ dulcis, et deinceps septentrionem versus stratum terræ fossilis (Tørv) inveniuntur. Intra præruptum loca demissiora, omnino lacus exhaustos referentia, et in ipso prærupto foramina, per quæ aqua in fretum diffluere potuit, reperiuntur. Paullo

---

1) Forchhammer: Studien am Meeresufer in Leonhards u. Bronns Neue Jahrb. für Miner u. Geogr. 1841 p. 10.

2) Cfr. Cateau-Catteville Beskrivelse af Østersøen, oversat af Rawert.

inferius ad Frederiksleie prope littus fossa fere 15′ supra superficiem maris manu facta erat, qua hæc strata nudata sunt (fig. 3): superne stratum humi, cum arena commixtæ, sesquipedale (a), deinde tenue stratum Molluscorum marinorum velut Cardii edulis, Littorinæ littoreæ, Paludinellæ Ulvæ, al. (b), tum stratum eodem genere, quo summum (a′), denique stratum saxorum erraticorum (Rullestene d.), sub quo argilla reperitur (e). Quæ formatio utrum diluvie, quæ æquor quindecim pedibus superaverit, an allevatione aliqua orta, putanda sit, decernere non audeo; illam autem explicationem verisimiliorem habeo. Qua re sola allevatio oræ Scaniæ hac periodo terræ significari videtur; vestigia autem demissionis nulla omnino invenire potui. Colligentes igitur mutationes, quas oræ circum fretum post formationem hac periodo terræ subierunt, has reperimus:

Fig. 3.

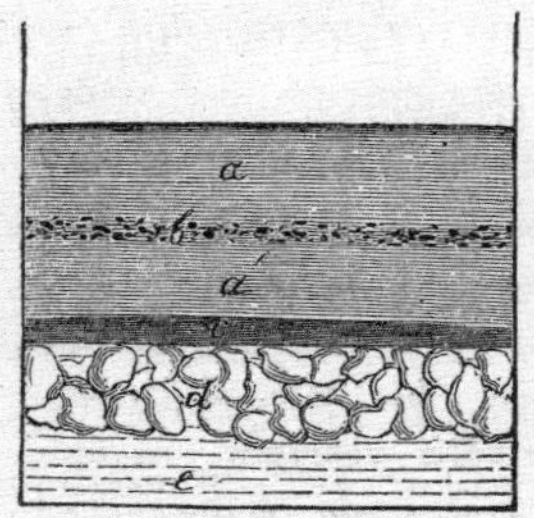

1) Neque Scanica neque Siællandica ora demissione mutata est.

2) Quæ res elevationem indicare videntur, in utraque ora parvulæ et eiusmodi sunt, ut decernere, annon eodem jure alluviis tribui possint, difficile sit.

3) Tota ora Siællandica, paucis locis, ubi prærupta est, exceptis, *alluvione* crevit, quod inprimis in australem et plus etiam in borealem convenit.

4) Pæne tota illa pars oræ Scanicæ, quæ freto adiacet, *abluvione* semper comminuitur; cui quantum spatium paullatim ademtum sit, non facile decerni potest.

5) Fretum igitur, postquam formatum est, tantummodo alluvione et abluvione, non elevatione aut demissione mutatum esse videtur.

Nota. Antequam descriptionem harum orarum relinquimus

universam naturam præcipitiorum argillaceorum, quamquam cum tota descriptione non necessario cohæret, contemplabimur. In ora enim Siællandica duo loca sunt, quibus præcipitia argillacea tam prærupta sunt, ut strata omnia nudentur, ad Vedbæk et inter Sletten et Humlebæk. Illic maxime ex argilla vulgari lutea præcipitium constat, modo superne stratum saxorum erraticorum maiorum, sed dispersorum et in medio stratum argillæ luteæ multo mollioris invenitur. Alterius præcipitii (fig. 4) major pars ex argilla vulgari constat (aa), sed hic tria reperiuntur strata densiora saxorum erraticorum maiorum minorumque (ccc), inter se et a reliqua argilla parvis stratis argillæ mollioris secreta (bbb). Ad imum præcipitium passim argilla glauca apparet, unoque loco stratum Calcarei conglomerati (d), tantæ extensionis, ut navibus Havniam portetur et cementi loco adhibeatur. Paululum in mare exit, quum argilla friabilis, qua tectum fuit, abluta sit. In ora Svecicaad Glumslöf præcipitium, ceteris Siællandicæ oræ et longius et altius, reperitur. (fig. 5). Satis magnum discrimen in diversis eius partibus apparet; hæc tamen generaliter afferenda sunt. Superior pars pæne dimidia ex vulgari argilla lutea arenosa (a) una cum multis saxis erraticis constat; superne duo triave strata calcis aquæ dulcis (ccc) et humi(bbb)alternant; inferior autem dimidia pars ex argilla constat fissili, glauca, plastica, cui omnino saxa erratica desunt. Notandæ vero sunt flexuræ multæ stratorum argillæ glaucæ

Fig. 4.

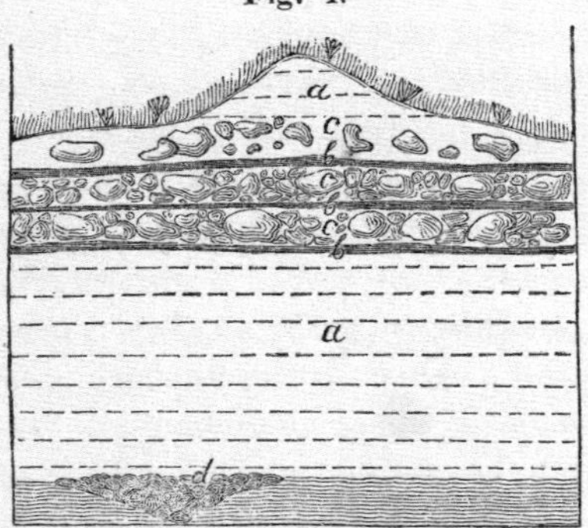

Fig. 5.

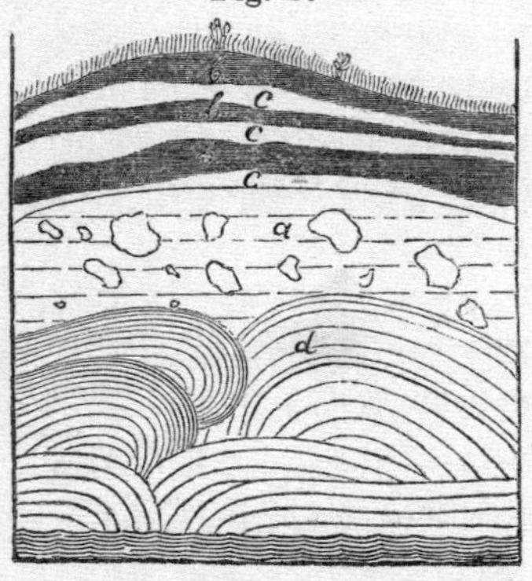

(d), quæ stratis flexis (blomkaalagtigt bøiede) prope Veile oppidum, a Forchhammero detectis [1]), omnino respondent. Vix dubitari potest, quin eadem sit fissilis plastica glauca argilla, quam Forchhammerus descripsit, quamque putat, ubi australis sinus Codani pars hodie sit, formationem magnam effecisse [2]). Ultimus hic est terminus, quem novimus, hujus formationis meridiem versus.

De natura fundi freti.

§ 9.

Descriptio fundi.

Supra vidimus, fretum fissuram esse vel potius planíssimam convallem, quæ maxima ex parte strata tertiaria, arenam et argillam, persecat. Quæ strata num per fundum freti eodem modo atque in terra continuentur, iam investigabimus. Facile reperiemus, hoc non ita esse, sed massas solutas fundi propriis legibus constitutas esse, ut *tria genera fundi existant, quippe qui 1) ex argilla, aut 2) ex lapidibus vel testis* Molluscorum *aut 3) ex arena constet.* Medium fretum *regio argillacea* [3]) diversissimæ latitudinis et distantiæ ab ora tenet. In boreali enim parte freti, apud Kullen, usque ad terram patet, a qua secundum totam oram Scanicam a Kullen ad Landskronam non procul abest; in Danica autem parte nullus fundus argillaceus prope oras nisi inter Hellebæk et Snedkersteen, vix spatio quartæ partis milliarii ab ora, reperitur. In reliquo freto fundus argillaceus vix obvenit nisi intervallo unius duorumve milliariorum ab ora. Prope omnes oras insulæ Hveen, quæ in medio freto emergit,

---

1) V. Kroyers Tidsskrift 1 Bd. P. 209.

2) L. c. P. 216.

3) In tabula colore luteo notata.

fundus argillaceus est. Hæc regio modo inter Helsingoram et Helsingburgiam, ubi fundus arenaceus est, abrumpitur. In utroque hujus zonæ latere reperitur angustissima zona, *regio lapidaria*[1]; ubi fundus lapidibus maioribus minoribusve tegitur. A Kullen ad Landskronam certis limitibus circumscripta est. In reliquo autem freto lapides magis dispersi sunt, ut hæc regio pæne evanescat. Ad Siællandiam modo per spatiolum meridiem versus ab Helsingora extra Snedkersteen et Espergjerde, et secundum littus occidentale insulæ Hveen certis limitibus circumdata est. Aliquot locis pro lapidibus magna multitudo testarum, imprimis Cyprinæ Islandicæ, reperitur, velut extra Hellebæk. Piscatores angustum spatium, ubi fundus ex testis congestis constat, „Skallekant" vocant.

Intra zonam lapidariam usque ad littus *regio arenaria*[1] patet. Fundus arenarius et lapidarius est, imprimis arenarius, quamquam lapides aliquot locis dominantur. Quod in majorem partem regionis, quæ septentrionem versus ab Helsingora et Helsingburgia patet, itemque in partem australem freti convenit. Extra Kullen solum hæc regio deest, quia argillacea littori adiacet. Ab Helsingora, ubi angustissima est, latitudine usque ad Kiöge, ubi totum spatium inter hanc urbem et Falsterbo occupat, semper sensim crescit. Idem in parte Suecica ab Helsingburgia ad Falsterbo fit. Circum Hveen, quamquam est angustissima, reperitur.

---

1) In tabula colore intense brunneo et rubro notata.

2) In tabula colore dilute brunneo, viridi, coeruleo, albo notata.

Explicatio formationis trium regionum geologicarum.

§ 10.

Ratio certa et definita, quæ inter fundi indolem et profunditatem semper apparet, leges, quibus ille formatus est, facile ostendit. Fundum igitur arenarium a littore ad certam profunditatem, 7—8 orgyia., patere, deinde paullatim magis argillaceum fieri videmus, ut pæne omnibus locis maioris profunditatis fundus argillaceus sit. Hic, igitur propius oram Scanicam quam Siællandicam invenitur, quod profunditas illi propior est quam huic, ut supra vidimus. — Si iam nullus alius motus maris esset, quam fluctuum, ita in fundo massæ solutæ dispertitæ essent, ut magnitudo particularum profunditati retrorsum responderet, ut igitur argilla mollissima in maxima profunditate, lapidesque maximi proxime littora reperirentur. Videbimus autem, hanc rationem alio motu maris modificari, scilicet qui efficitur flumine; quod tertiam regionem lapidariam constituit. Facile enim observatur, eam imprimis esse, ubi vis fluminis propter configurationem littorum violentior est, itaque abluvione arenæ argillæque, qua lapides soli relinquuntur, effici. Itaque vehemens flumen angustissimæ partis freti inter Helsingoram et Helsingburgiam efficit, ut hic argilla in profunditate solita non inveniatur.

Statuendum est igitur, quum aqua in convallem, quæ tunc Siællandiam a Suecia secernebat, primum irrumperet, strata fundi huius convallis eodem modo, quo in oris fretum cingentibus hodie invenimus, ordinata fuisse; arenam igitur argillamque inter se commixtas fuisse. Nunc vero aliter ordinata esse, quia duæ rationes, vis aquæ motoria (principium mutans movensque) et profunditas (principium sedans mutationique resistens), in contrarium nitantur. Isto enim tempore argillam mollem minimasque arenæ particulas omnibus locis, quibus profunditas tam parva esset, ut motus undarum fundum quoque afficeret, ab ora ad profunditatem orgyiarum 7—8, aqua turbatas esse, deinde eas freti partes, quarum profunditas tanta esset, ut eas motus aquæ afficere non posset, i. e. locis 7—8 orgyiis profundioribus, petiisse. Ita facile intelligitur, regionem argillaceam loca profundissima freti complentem, utrinque regione arenaria circumdatam esse. Item ubicunque fundus assurgit, arena appareat, necesse est. Itaque omnia vada arenaria sunt et, velut insulæ, zona argillacea cincta. Quod ita quoque etiam dicere possumus: in freto omnia superiora loca, vada, i. e. colles maris, ex argilla constare. Vado longe patenti, quod austrum versus ab Helsingora situm Diſken vocatur, hæc sententia affirmatur.

Et vim motus et profunditatem spectandas esse, perspicuum est, si formatio, quæ Marſk vocatur, oræ occidentalis Slesviciæ observetur. Ostendit enim, in parvula profunditate argillam, si motus quoque parvus sit (quod in serie insularum ad littus occidentale Slesviciæ apparet), in maiore autem arenam, si violentior maris motus sit (velut extra illas insulas), secerni posse.

Ex hac igitur paragrapho haec statuimus.

1) Fundi maris, si ex massis solutis constat, tria sola genera sunt: 1) arena, 2) argilla, 3) lapides vel testæ congestæ.

2) Quodnam genus fundi ex hisce tribus generibus in mari appareat, in ratione profunditatis ad vim undarum et in flumine positum est.

3) Ubicunque profunditas non major est, quam ut fundus motu undarum affici possit (in freto ab ora usque ad profunditatem 7—8 orgyiar.), arena apparet, argilla, cujus particulæ multo minores sunt, abluta.

4) Ubicunque profunditas major est, quam ut fundus

motu undarum affici possit (in freto a profunditate 7—8 orgyiar.) argilla, quæ in locis maris minus profundis abluta hic demissa est, apparet

5) Fundus igitur arenarius littora aut vada, argillaceus profunditatem indicat.

6) Fundus lapidarius vel testaceus, loca maris, quorum flumen vehementissimum est, indicat.

7) Hæc igitur distributio massarum modo phænomenon superficiei habenda est.

Resultata prioris paragraphi in geognosiam translata.

§ 11.

Non credo, adhuc satis intellectum esse, quantum ad geologiam illustrandam pervestigatio naturæ fundi valeat; quamquam perspicuum erit partim si reputaverimus, formationes plerasque geologicas, quæ normales vocantur, in fundo maris exstitisse, partim si meminerimus, opinionem illorum, formationes antiquas consentienter iis, quæ hac terræ periodo fiunt, expediri posse statuentium, semper magis magisque confirmari. Quamquam scientia pervestigationibus cl. Forchhammeri, quomodo mare littora afficiat, magnopere aucta est[1]), in sequentibus tamen videbimus, inprimis si comparabimus, quomodo mare in diversa profunditate diverso modo fundum afficiat, tum demum apparere, quantum examina hæc valeant. In superioribus unitatem varietatis fundi demonstrare conati sumus, modo tria genera fundi esse probantes, et causas harum varieta-

---

[1]) Leonhard und Bronn, Neues Jahrbuch für Mineralogie und Geog. loc. cit.

tum fundi, per diversas periodos terræ, vel, quod idem est, in formationum normalium stratis eandem unitatem esse probare possumus, easdem vel analogas diversitates causarum valuisse, nobis coniicere licebit.

Fundus freti cum formatione tertiaria Daniæ comparatus.

§ 12.

Formationes, ad tertiariam pertinentes, ex argilla arenaque constantes, pæne totam superficiem Daniæ tegunt. A cl. Forchhammero in tres partes distributæ sunt: 1) formationem electricarbonis (Ravbrunkulformation), 2) argillam tertiariam (Rullesteensleer), 3) arenam tertiariam (Rullesteenssand), quas et tempore et modo diverso ortas esse putat[1]). Mihi autem, fundum freti, perscrutato, dubitatio quædam exstitit, num membra duo posteriora ad diversum formationis tempus referri possent, et num eo modo, quo putat, eæ ortæ essent. Libenter quidem confiteor, me non sine pudore quodam opiniones a geologo egregio dissentientes proferre audere. Ut ego, si sententia mea refellatur, eam revocare paratus sum, sic libentisssime, si quid in his recte sentiam, doctissimo illi viro, cuius secundum præcepta investigationes meas institui, deberi agnosco. Respicientes enim, quo modo illæ duæ partes formationis distributæ sint, arenam in altissimis locis et omnibus collibus maioribus, argillam modo iu vallibus reperiemus[2]). Quæ distributio diversa rationem diversam,

---

1) V. Krøyers Tidsskrift 3 B. P. 546.

2) V. Geognostisk Kort over Danmark af Forchhammer 1843.

qua formatæ sint, perspicue indicare mihi videtur. Si enim fundus freti super æquor allevaretur, minorem imaginem rationum, quas duæ partes formationis tertiariæ maiore modo præbent, videremus. Omnia enim loca altiora, quæ antea vada maris fuerunt, ex arena, omnes convalles, antea profunditates, ex argilla constarent. Quas deinde rationes cum Daniæ formatione tertiaria, quæ certe antea fundum maris, quo Dania tecta erat, similiter formavit, comparando, necesse esse facile intelligimus, eam formationis partem, quæ arena tertiaria vocata hodie clivos efficit, vada maris, eam autem partem, quæ argilla vocata planities hodie complet, profunditates maris maiores, fuisse; has igitur formationes non diverso tempore, sed æquales sub diversis eiusdem maris condicionibus exstitisse.

Formatio quoque regionum Daniæ, quæ æque humiles atque argilla tertiaria sunt, quarumque solum ex arena constat, ex antecedentibus explicari potest; sub iisdem enim condicionibus, quibus pars angustissima freti, ubi arenam quoque in profunditate alias argillam præbente reperimus, formatæ existimandæ sunt. (Vide tabulam geographicam inter Helsingoram et Helsingburgiam).

Nota 1. Veritati huius sententiæ de formatione tertiaria Daniæ fortasse obiicietur, quod illæ duæ partes fossilia diversa habent, vel potius, quod alteri, argillæ, fossilia omnino desunt, altera, arena, reliquias continet animalium, quæ hodie quoque maria nostra habitant [1]; hinc igitur has formationes diverso tempore exstitisse coniici certe licet. Primum autem notandum est, quamquam fossilia in argilla nondum reperta sunt, tamen fieri posse, ut postea reperiantur, quod eo verisimilius fit, quod nulla fossilia reperientes maximam partem arenæ, rimari possumus, quia in hac formatione illa rarissima sunt. Sed si intelligatur, certum esse, aut diversa fossilia in iis contineri, aut fossilia modo in altera sola esse, eiusdem tamen temporis formationes eas esse posse credo. Nam

---

1) V. Krøyers Tidsskrift, 3 B. p. 548.

quoque in hoc mari animalia fundi arenacei ab animalibus fundi argillacei valde distant, ut in sequentibus demonstrabimus. Fortasse quoque objicietur, quod ipse in superioribus notavi, hanc distributionem massarum solutarum fundi modo in superficiem convenire (v. § 10); vada igitur maris, atque etiam colles terræ, qui ad arenam tertiariam pertinent, si analogia vera est, in sola superficie ex arena constantia nucleum argillæ includere. Ad hæc respondeo, me, donec contraria probentur, hanc rationem æque in vada maris ac in colles convenire, putare. Denique notandum est, si difficultates in sententia a me proposita appareant, quas nunc tollere nequeam, me tamen non intelligere, quomodo fieri non possit, quin tales analogas rationes inter naturam huius fundi ejusque, qui tum erat, existere, necesse sit.

Nota 2. Alias sententias eiusdem auctoris egregii de geologia Daniæ resultatis, quæ effectus maris in fundo observando adeptus sum, accommodare non possum. Insulæ enim, quæ in parte freti australi occidentem versus reperiuntur, argillaceæ putantur, quod diluvies, quæ ab oriente fluxisse totamque terram inundasse existimatur, arenam abluit[1]. Ex antecedentibus facile intelligitur, mare violenter agitatum arenam abluere non posse, nisi simul abluatur argilla, quae, ex particulis multo minoribus, quæ facilius motu maris agitantur, constans facilius etiam aufertur. Itaque talis diluvies potius effecisse nobis videtur, ut argilla, arena remanente, ablueretur.

Quod alio loco[2] ille geognostes clarissimus putat, sententitiam suam stratum argillæ inter formationem arenæ viridis (Grønsand) et Calcareum insulæ Saltholmiæ esse, eo confirmari, quod argilla ad septentrionem a Saltholmia in profunditate pedum septuaginta duorum reperiatur, ex antecedentibus scimus, ubique in freto in hac profunditate argillam inveniri, et quidem modo in

---

[1]) Skandinaviens geognostiske Natur, et Foredrag holdt d. 22 Novbr. 1843 af Forchhammer p. 18.

[2]) Danmarks geognostiske Forhold af Forchhammer. Kjøbhvn. 1835 p. 49.

superficie, ut inde nihil certi de stratorum natura in maiore profunditate conicii possit.

Ratio inter indolem fundi marini et geognosiam universam.

§ 13.

Quamquam antea indicatum est, saxa in omnibus formationibus, quæ normales vocantur, ad tria genera: 1) Schistum argillaceum, 2) Arenarium, 3) Calcareum referri posse, adhuc tamen interior coniunctio harum trium partium earumque ratio ad membra omnino analoga massarum solutarum, quas maria hodie quoque hac terræ periodo habent, vix satis animadversa est. Vidimus enim, mare secundum rationes motu maris, profunditate, flumine constitutas, arenam aut argillam aut calcem secernere. Iam analoga membra formationum, omnibus probatarum, ostendere enitemur.

Condicio uniuscujusque formationis normalis est terra allevata saxorum primitivorum. Horum enim semper ex massis, quæ violentibus frequentibusque allevationibus et demissionibus divulsae sunt, inter certos maris cingentis fines formationes conficiebantur normales i. e. quæ, per strata in mari deposita, reliquias organismorum continent. Quum materies primitiva eadem semper fuerit, scilicet granitus, hinc perspicimus, cur saxa omnium formationum ad analoga membra referri possint. Granito enim et gneisso soluto, argillam arenamque formari scimus [1]), si-

---

[1]) Forchhammer om Leerarternes Oprindelse i Vidensk. Selskabs Skrifter 1832.

mulque calcem ex interiore terra primitus erupisse. Ad hoc minus interest, utrum ista saxa Schisto argillaceo an argilla soluta, utrum Arenario an arena soluta, utrum Calcareo an Creta confecta sint; qua re solæ modificationes materiei pæne eiusdem, externis effectibus exortæ, indicantur. In sequenti conspectu exponentur analoga membra parallelarum serierum, quæ formationes diversas constituunt, ut etiam in geologia affirmetur sententia, quæ in distributione animalium plantarumque systematica valet, omnem evolutionem in natura per parallelas series analogorum membrorum fieri [1]).

Conspectus membronum analogorum omnium formatiorum.

I. Formatio siluriana.

1. Arenarius argillaceus (Graawacke).	2. Schistus argillaceus.	3. Calcareus transitorius.

II. Formatio carbonifera.

1. Arenarius carboniferus.	2. Schistus carboniferus.	3. Calcareus carboniferus.

III. Formatio Arenarei rubri.

1. Arenarius ruber.	2. Schistus cupreus.	3. Calcareus alpinus. (Zechsteen).

---

1) Vide: Havetidenden 9de Aarg. 5te og 11te Hefte, et Entwurf einer systematischen Eint. der Platwürmer von Ørsted, p. 33 1844.

IV. Formatio Arenarei variegati.

1. Arenarius variegatus.	2. Argilla (Marga) variegata. (Keuper).	3. Calcareus Conchifer.

V. Formatio Jurassiana.

1. Arenarius Liasianus.	2. Schistus. Liasianus.	3. Calcareus Jurassianus.

VI. Formatio Cretæ

1. Arena viridis.	2. Argilla?	3. Creta.

VII. Formatio tertiaria.

1. Arena tertiaria.	2. Argilla tertiaria.	3. Calcareus?

In his igitur seriebus formationes, quæ 1^0 notatæ sunt, arenæ respondent, quæ 2^0 notatæ sunt, argillæ, quæque 3^0, testis Molluscorum congestis, et, si analogia iusta est, omnes eodem modo ortæ sunt.

Tria illa membra in omnibus formationibus non æque regnant; in antiquissima enim argilla multum valet, quum non solum Schistus argillaceus his multo maioris extensionis quam in ceteris formationibus sit, sed etiam membro, arenæ respondenti, multum argillæ admixtum sit, quia Grauvacke est Arenarius argilla conglutinatus. Deinde formationes arenariæ frequentiores esse incipiunt, donec formatione, quæ Arenarius variegatus vocatur, ad summum evolutionis gradum perveniunt, ut membrum, quod argillæ respondet, magnopere recedat et minus distincte in formatione Keuperiana, quæ tamen maximam partem ex argilla vel marga constat, apparet. Post hanc formationem tertium membrum, Calcareus, usque ad formationem tertiariam plurimum valet. Itaque forma tiones omnes, tertiaria vetustiores, ad tres sectiones, quatenus arena aut argilla aut calcareus dominatur, referri possunt.

1 Sect. Argillæ.

Format. Grauvackes.

2 Sect. Arenæ.

Format. carbonifera; form. Arenarei rubri; form. Arenarei variegati.

3 Sect. Calcis.

Format. Jurassiana, form. Cretæ.

Quo tempore Grauvacke allevata est, plana littora non erant, quæ arenam non procreare non potuerunt. Mare profundum insulas cinxit, ut fundus, qui allevabatur, pæne ex argilla sola constaret. Formationes, quæ deinceps usque ad formationem Jurassianam sequuntur, magna littorum plana indicant. Formationes Arenarei et in Scandinavia et in Bornholmia cum littoribus arcte coniunctas esse, cl. Forchhammerus iam probavit[1]). Hæc littora silvis tecta erant, quæ, demissione submarinæ factæ, arena oppletæ in strata Lithanthracis mutatæ sunt; quæ phænomena omnino iis similia sunt, quæ in littore occidentali Iutlandiæ multo recentiore tempore facta esse, idem cl. geologus probavit[2]); hic vero silvæ terram fossilem efficiunt, nec arena in Arenarium mutata est. Item huius periodi formationes salis fossilis frequentes phænomena indicant, iis, quæ nunc in planis salinis Russiæ australis fiunt, simillima.

Periodo calcaria rationes fuisse necesse est, quæ in-

---

1) Falck: Statsbürgerliches Magazin 7 Band. Videnskabernes Selskabets Skrifter 7 Deel p. 19.

2) Oversigt over Videnskabernes Selskabes Forhandlinger i 1842 p. 64.

primis animalibus testaceis faverent; strata enim, eorum reliquiis confecta, hic multo maioris extensionis quam strata argillacea et arenacea sunt; unde a viribus organicis mechanicas, quæ antea plurimum valuerant, superatas esse intelligitur.

CAPUT SECUNDUM.

De regionibus Algarum in freto Öresundico.

§ 14.

Omnibus, qui ut ego studium algologicum ab algis freti investigandis incipiunt, mirum videtur, quod scripta, in quibus de algis septentrionalibus agitur, auctores mare, quod iis proximum esset, nimis neglexisse declarant. Hydrophytologia enim Danica Lyngbyei, quod ad maria Danica pertinet, inprimis examinationibus sinus Othiniensis nititur, neque in scriptis algologicis cl. Agardhii distributio algarum in hoc mari specialius illustratur, ut inde generales leges appareant. Hornemannus hac in re alios sequitur. Graviora autem adiumenta recentioribus temporibus ab J. Agardhio[1]) et Liebmanno[2]) allata sunt. Præcipue illius investigationes maximi momenti sunt, quod ex iis leges de distributione algarum universa verticali exeunt; quæ breviter hæ sunt:

„Secundum distributionem geographicam Algæ Scandinavicæ in tria regna subordinibus tribus Algarum respondentia abeunt:

1) *Regnum Algarum Zoospermarum*, totam vegetationem aquæ dulcis et marinam magis amphibiam complectitur, unde culmen huius regni, si tantum thalassiophytas respicias, ad littora nostra orientalia positum esse existimandum est. Hoc regnum in regiones duas dividendum est.

---

1) Novitæ floræ Sûeciæ ex Algarum familia. Lundæ 1835.

2) Krøyers Tidsskrift 2 B. p. 464.

a) *Regio Confervarum*, complectens Algas aquæ dulcis.

b) *Regio Ulvacearum*. In hac regione dominantur Ulvæ, Conf. ærea et C. rupestris.

2) *Regnum Algarum Olivacearum*, inter Zoospermas et Florideas quodammodo intermediæ. Algæ olivaceæ maria salsiora præoptant, licet ibi ad sinus magis reclusos confugiant, neque maria minus salsa respuant, formis attamen contractis in his provenientes. Iis sequentes regiones assumere licet:

a) *Regio Lichinæ*.

b) *Regio Sphacellariæarum* in scrobiculis inferioribus mare sæpe accipientibus.

c) *Regio Fucorum* in ipso limine maris.

d) *Regio Dictyotearum* in fundis 3—6 orgyias infra limitem maris demersis.

e) *Regio Chordariæarum* in rupibus æstui maris magis expositis.

3) *Regnum Algarum Floridearum* in aperto mari, profunditate circiter 6—14 orgyiarum.

a) *Regio Condriæarum* in fundis a mari non nimis reclusis.

b) *Regio Delesseriæarum* in profunditate 9—10 orgyiarum."

Quamquam, ut ex sequentibus intelligetur, cum algologo clarissimo in summis rebus consentimus, quod Algæ secundum distributionem in tria regna, quæ tribus sectionibus earum respondent, dividendæ sunt, in hac tamen auctoris divisione distributionem horizontalem et verticalem non satis distinctas esse putamus. Quod ne ego quidem in conspectu distributionis plantarum per fretum, quem antea attuli, observavi[1]. Animadvertendum enim

---

[1] Förhandlingar vid de Skandinaviske Naturforskarnes tredje Møte i Stockholm 1842 p. 621.

est, distributionem Algarum a littore profunditatem versus distributioni plantarum a summo ad imum montem respondere, itâque directionem verticalem indicare. Hac de causa divisiones, non regna, ut apud illum auctorem, sed regiones nominandæ sunt. Itaque in sequentibus distributio algarum per fretum et verticalis et horizontalis examinabitur. Apud illum auctorem desunt quoque observationes, quomodo distributio sit in profundioribus maris locis.

§ 15.

In fundo freti ab ora Siællandica ad Scanicam proficiscentes, quæ diversæ regiones plantarum in utroque convallis latere, quæ nobis transeunda est, sibi succedant, videbimus. Initio præcipue Oscillatoriæ, Confervæ, Ulvæ, omnes igitur algæ virides (Chlorospermeæ) obveniunt. Quibus in profunditate orgyiarum 2—5 Fucoideæ et in paullo maiore profunditate Laminarieæ, ergo omnes algæ colore olivaceo distinctæ, Melanospermeæ, succedere incipiunt. In profunditate orgyiarum 8—10 algæ olivaceæ sensim evanescunt, et purpureæ (Chondriæ, Delesseriæ, Ceramieæ) succedunt. Hæ sunt igitur illæ plantæ, quæ in summa profunditate reperiuntur; sed per medium fretum zona, vegetatione omni carens, totum fere spatium, quod supra nomine zonæ argillaceæ descriptum est, patet. Tres illæ regiones in utroque latere freti, sed in diversis locis diversa extensione obveniunt. Quæ diversitates tabulas, quarum altera (Tab. I) regiones supra, altera (Tab. II) in proiectione verticali conspectas adumbrat, leviter aspiciendo facile animadvertuntur. Continuatio regionum plantarum, quæ in montibus inveniuntur, haberi possunt; quare regiones nominandæ sunt. Iam singulas regiones diligentius examinemus.

§ 16.

Regio Algarum viridium s. Chlorospermearum.

(In tabula colore coeruleo et viridi notata).

Hæc regio ex ipsa ora ad profunditatem orgyiarum 2—5 patens, algis viridibus, Oscillatoriis, Confervis, Ulvis, distincta est. Nulli loco freti omnino deest, sed diversissima extensione est. In parte freti australi maxima est extensione, et septentrionem versus sensim minore, quod inde fit, quia pars australis eadem fere indole est, ac lacus, in quibus algæ virides proprie dominantur. Hæc regio in duas subregiones: 1) subreg. *Oscillatorinearum s. Algarum cœruleoviridium,* 2) *subr. Ulvacearum* dividi potest.

Subregio Oscillatorinearum. Hæc subregio ea pars regionis algarum viridium est, quæ oræ adiacet et frequentissime aqua omnino denudatur. Iam investigationes ad Hofmansgave a me antea institutæ probarunt, Oscillatorineas in littore plano proximo terræ dominari[1]). Et in freto Oscillatoriæ inter omnes algas superficiei maris maxime appropinquant. Ita in vadis subsalsis i. e. in stagnis prope littus, quæ fluctu recedente coniunctionem nullam cum mari habent, *Lyngbyea glutinosa, L. æstuarii, Spirulina subsalsa, Microcoleus chthonoplastes, M. fuscus* obveniunt. Plantæ horum locorum propriæ nominandæ hæ quoque sunt: *Conferva fracta, C. perreptans, C. Linnum, Ulothrix floccosa, Ulva clathrata, Erythroconis littoralis.* In lapidibus ad munimentum Trekroner in ipso limite maris *Lyngbyea lutescens,* et ad Kullaberg *Callothrix fasciculata* obvia. Hæc tam fre-

---

1) Krøyers Tidsskrift Bd. 3 p. 552. Beretning om en Excursion til Trindelen i Odensefjord af Ørsted.

quens in saxis invenitur, ut ea, mari recedente, quasi nigro colore obducta videantur.

Subregio Ulvacearum. Hæc subregio partem regionis algarum viridium, quæ aut raro aut nunquam aqua denudatur, occupat. Facile intelligitur, hic, ut in talibus divisionibus, nullos certos terminos indicari posse; nam semper quædam plantæ eodem iure ad plures divisiones referri possunt. Ex algis viridibus, quæ superficiei maris tam appropinquant, ut dubium sit, num ad priorem subregionem referendæ sint, nominandæ: *Ulva intestinalis, compressa, Linza, Conferva ærea, uncialis, vaucheriæformis. Hormiscia penicilliformis, H. assimilis, Ulothrix contorta, Cruoria pellita, aliæque.* Huius subregionis propriæ sunt: *Ulva lactuca et latissima,* quibus præcipue sinus cum fundo limoso abundant, velut in Kallebodstrand, inter Havniam et Trekroner, ad Landskronam, al. Interdum in profunditate orgyiarum 3—5 obveniunt. *Conferva rupestris, distans, glomerata, gracilis, elegans* in profunditate orgyiarum 6—7 occurrunt.

Nota. In hac regione quanquam algæ virides ita dominantur, ut eæ solæ vegetationem propriam efficere dici possint, nonnullæ tamen algæ olivaceæ et purpureæ reperiuntur; sed nullam algam hujus regionis purpuream, cuius color pulcher est, nominare possum. Ex illis notandæ sunt: Lichina confinis, Ilea fascia, Chorda lomentaria, Ectocarpus littoralis, siliculosus, tomentosus, Ceramium diaphanum, Hutchinsia nigrescens.

---

§ 17.

Enumeratio Algarum viridium s. Chlorospermearum.

Conferveæ

Conferva centralis Lgb. ad Kullaberg. — *Conferva globosa* Ag. in stagnis marinis insulæ Gråen ad Landskronam Ag. — *Conferva uncialis* Lgb. Kullen-Trekroner. — *Conferva sericea* Ag. Kullen-Trekroner. — *Conferva refracta* Ag. (Conferva fracta marina Lgb.). — *Conferva glomerata* L. — *Conferva rupestris* L. — *Conferva arcta* Dillv. Kullen — Trekroner. — *Conferva vaucheriæformis* Ag. Trekroner. — *Conferva gracilis* Griff. Kullen-Hveen. — *Conferva elegans* nob. inter Taarbæk et Hveen[1]) — *Conferva distans* Dillv. (Conf. Hutchinsiæ Flor. Dan. t. 2 314) inter Havniam et Trekroner. — *Conferva obtusangula* Lgb. Taarbæk. — *Conferva implexa* Dillv. Helsingor. Liebm. Trekroner. — *Conferva Linum* Roth. — *Conferva perreptans* Carm. (Zygnema littoreum Lgb.) ad Havniam (Ny Badehuus) — *Conferva rigida* Ag. prope Landskronam ad Gråen Ag. — *Conferva Melagonium* Web. et M., ad Helsing. Liebm. — *Conferva ærea* Dillv. ad Hellebæk. — *Conferva Hofmanni* Ag. Trekroner Ag. — *Bombycina marina* nob. (Conferva bombycina submarina?) ad Havniam (Kallebodstrand ad ny Badehuus[2]). — *Hormiscia assimilis* nob. ad Havniam (in lapidibus ad Langelinie). — *Hormiscia penicilliformis Fr.* (Conferva hormoides Lgb). Kullen, Helsingor Trekroner[3]). — *Hormiscia flacca* nob. (Myxo-

---

[1]) Hæc elegantissima species a Conf. fracta, cui proxima est, præcipue articulorum infima parte ramis adnata differt.

[2]) Differt a Conf. bombycina (Bombycina stagnalis nob.) articulis duplo brevioribus, geniculis valde constrictis. Hæ duæ species genus distinctissimum, præcipue massæ sporaceæ indole ab omnibus aliis recedens, constituunt. Distingui potest hoc modo: Fila simplicia fluctuantia, mucosa; genicula distincta plus minusve constricta, ut striæ simplices pellucidæ apparentia; singuli articuli sporidia duas massas discretas, geniculis adpressas et inter se tantummodo filis tenuissimis coniunctas, formantia.

[3]) Hormisciæ generis nota, ut puto, gravissima, adhuc omnino

nema flaccum Liebm.[1]) Kroyers Tidsskrift 2 B. p. 323). *Conferva flacca* Dillv. ad Hornbæk et Helsingor. Liebm. — *Ceramicola*[2]) *rubra nob.* (Conferva ceramicola Lgb., Callithamnium ceramicola Suhr., Ceramium ceramicola Ag.) in Ceramio rubro ad Hofmansgave. — *Ulothrix subsalsa* nob. ad Havniam (in Kallebodstrand ad Ny

---

neglecta est; hanc enim in eo constare puto, quod massa sporacea zona pellucida circumdata sit, ita ut tubus duplex adesse videatur, et interior et exterior. Hoc modo fili pellucidæ partis insolita magnitudo oritur. Est fortasse hic character, quem cl. Friesius proposuit verbis: „Articuli turgent in fila pulposa gelatinosa.“ Tres antice nominatæ species distingui possunt hoc modo:

Hormiscia assimilis. Articulis ferme duplo longioribus quam latis, geniculis parumper constrictis.

Hormiscia penicilliformis. Articulis aeque longis ac latis, geniculis valde constrictis.

Hormiscia flacca. Articulis duplo latioribus quam longis, geniculis parum constrictis. Præterea differunt inter se magnitudine diversa tubi pellucidi interni.

1) Ad genus Myxonematis Fr. plantæ diversissimæ affinitatis relatæ sunt. Si Confervam lubricam pro generis typo habes, omnes ceteræ species, quæ non ramosæ plura sporidia in quoque articulo continent, ab eo longe removendæ sunt. Ex his Conferva flacca Dillv. omnino cum Hormiscia characteribus genericis congruens cum hoc genere coniungi potest. Nonne Myxonema curvatum Liebm. Scytonematis species est? Conferva lubrica genus distinctissimum et ramificationibus et præcipue præsentia sporidii unici in singulis articulis constituit; inter Confervas ramosas analogon ad Ulothricis genus inter simplices.

2) Character huius generis distinctissimi: Fila simplicia affixa cespitosa, genicula distinctissima constricta ut striæ duplices apparentia, unicum sporidium in singulo articulo. Ab Ulothricis genere, cui proximum, differt geniculis multo distinctioribus et filis cespitosis; est idem ad Ulothricem, quod Callothrix ad Lyngbyeam.

Badehuus)[1]. — *Ulothrix* (Bispora) *floccosa* nob. (Conf. floccosa Ag.) ad Havniam in Kallebodstrand, ad zosteram marinam[2]. —

Siphoneæ.

Bryopsis arbuscula Ag. inter Taarbæk et Hveen. —

Ulvaceæ.

Ulva Lactuca L. — *Ulva latissima* L. — *Ulva Linza* L. — *Ulva compressa* L. — *Enteromorpha clathrata* Lk. — *Enteromorpha* (Scytosiphon Lgb.) *erecta* (Solenia clathrata confervoidea Ag.). *Erythroconis littoralis* Örsd. Kroyers Tidsskrift 3 B. Tab. 7 f. 1—3.

Oscillatorineæ.

Rivularia atra Roth. — *Rivularia pellucida* Ag. ad plantas marinas freti sundici Ag., mihi ignota. — *Rivularia nitida* Ag. ad Trekroner. — *Rivularia* (Scytochloria) *plana* Harv. in Hook Brit. flor. V. 2 p. 394 ad Hveen. — *Callothrix scopulorum* Ag. in Cerameis ad Kullaberg. — *Callothrix fasciculata* in saxis ad Kullaberg. — *Lyngbyea glutinosa* Ag. (Oscillatoria maiuscula Lyb.) — *Lyngbyea æstuarii* Liebm. (Lyngbyea ferruginea Ag. Oscillatoria æstuarii Lgb.) — *Lyngbyea crispa* Ag. Amager. — *Lyngbyea lutescens* Liebm. Trekroner. — *Microcoleus fuscus* Örsd. ad Havniam (ad Ny Badehuus). — *Spirulina subsalsa* Örsd.[3]) cum præce-

1) Distinguitur filis flexuoso- contortis cespitem minutissimum formantibus, articulis diametro æqualibus vel sesqui longioribus, in medio singulo sporidio unico ovali articulo, maculis duobus triangularibus decoloribus circumdato.

2) Hoc subgenus distinguitur: massa sporacea in duobus sporidiis terminata; in Ulothricibus veris unicum tantum sporidium in medio articulo. Conferva floccosa a cl. Agardhio aquæ dulci propria indicatur; sed secundum specimina ab Hofman Bangio ad me missa et ab hoc peritissimo algologo in mari inventa est.

3) Kroyers Tidsskrift 3 B. p. 566.

dente. — *Oscillatoria lutea* Ag. in saxis ad litus prope urbem Helsingburgiam in ipso limite maris Ag. ad Havniam. — *Leucothrix mucor* nob.[1]) in stagnis submarinis plantas filis suis radiantibus mucorum modo obducit.

Diatomaceæ.

Gallionella lineata Ehrb. (Fragillaria lineata Lgb.) ad Trekroner, Havniam (Ny Badehuus). — *Gallionella moniliformis* Ehrb. (Fragillaria numuloides Lgb.) ibidem. — *Agonium centrale* nob. ad Trekroner.[2]) — *Ceratoneis Closterium* Ehrb. (Act. Acad. Berol. 1840). — *Navicula Baltica* Ehrb. — *Navicula sigma* Ehrb. — *Navicula lanceolata* Ehrb. — *Navicula bifrons* Ehrb. — *Sigmatella Nitzschii* Kütz. — *Amphiprora constricta* Ehrb. Act. Acad. Berol. 1841 p. 410. — *Eunotia turgida* Ehrb. — *Eunotia Westermanni* Ehrb. — *Eunotia Faba* Ehrb. — *Cocconeis Scutellum* Ehrb. — *Cocconeis te-*

---

1) Character genericus. Fila alba et cæspites et stratum, radiis longissimis emissis, formantia, articuli distincti, interstitiæ hyalinæ simplicissimæ absque constrictionibus, sporidia numerosa pulveracea alba.

Hoc genus a Callothrice, cui proximum est, et indole cæspitis, valde radiantis, et interstitiis multo distinctioribus, ab omnibus confinibus sporidiorum indole valde recedit. Motiunculæ ut apud ceteras Oscillatorineas interdum conspici possunt.

2) Character genericus. Fila tenuissima rigidiuscula subflexuosa, cæspites formantia, distincte articulata, sed absque constrictionibus, instertitiæ rectæ hyalinæ, sporidium unicum in singulo articulo.

Absentia constrictionum et præsentia sporidii unici a confinibus valde recedit et quodammodo transitum ad Oscillatorineas constituit.

Agonium centrale. Filis pellucidis 3—5‴ longis a centro communi exeuntibus lapidibus affixis, articulis duplo latioribus quam longis, sporidiis ovalibus lutescentibus.

nuis nob. Surirella tenuis. Dujardin [1]. Num distincta species? — *Homalodiscus vulgaris* nob. vulgatissimus in plerisque plantis marinis [2]. — *Homalodiscus ovalis* nob. cum præcedente. — *Bacillaria vulgaris* Ehrb. — *Bacillaria elongata* Ehrb. — *Bacillaria flocculosa* Ehrb. *Diatoma marinum* Lgb. Bacillaria adriatica. Hyac. v. Lobarzeuski Linnea V. 14. Diatoma signata, sera et cælata [3], Grammatophora mexicana, stricta et oceanica Erb. [4] — *Striatella unipunctata* Ag. — *Odentella aurita* Ag. — Tessella catena Ehrb. — *Tessella arcuata* Ehrb. — *Tessella interrupta* Ehrb. — *Tessella Ralesii* nob. (Tessella catena Rales, Jardines. Mag. of nat. hist. Aug. 1840. Pl. 2 f. 1. — *Synedra ulna* Ehrb. (Echinella obtusa Lgb.) — *Synedra capitata* Ehrb. — *Synedra Gallionei* (Echinella fasciculata Lgb.) — *Podosphenia gracilis* Ehrb. — *Podosphenia abbreviata* Ehrb. — *Podosphenia cuneata* Ehrb. — *Achnanthes longipes* Ag. in palis ad Toldboden prope Havniam. — *Achnanthes brevipes* Ag. — *Achnanthes subsessilis* Kütz. — *Achnanthes minutissima* Ehrb. — *Schizonema rutilans* Ag. (Bangia rutilans Lgb., Naunema Hofmanni Ehrb.) ad Havniam (Toldboden in palis, Ny Badehuus). — *Acineta tuberosa* Ehrb. (Ny Badehuus).

---

1) Observat. au Microsc. Pl. 13. fig. 14.

2) Charact. gener. Cellulæ disciformes costis et lineis omnino destitutæ, sporidia 5- 6, e centro communi radiantia. *Homalodiscus vulgaris*, cellulis rotundis, sporidiis 5—6. *Homalodiscus ovalis*, cellulis ovalibus, sporidiis indistinctis.

3) Dujardin loc. cit. Pl. 30 f. 20.

4) Act. Acad. Berol. 1841. p. 378.

§ 18.

Regio Algarum olivacearum s. Melanospermearum.

(In tabula colore et dilute et intense bruneo notata).

Regionem algarum olivacearum extra et profundius, quam viridium in profunditate a 3—5 et ad 7—8 orgyias sitam esse, iam supra vidimus. Tabulam I, in qua hæc regio colore brunneo notata est, contemplanti facile apparet, eam maxima extensione in partibus latissimis freti esse, nec ullo loco abesse; hic enim maris profunditas his plantis idonea est, et fundus ex arena lapidibus maioribus minoribusve commixta constat. Hæc regio in duas subregiones dividi potest, a) subregionem Focoidearum et zosteræ marinæ, b) subr. Laminariearum.

Subregio Fucoidearum et Zosteræ marinæ.

(In tabula colore dilute brunneo notota).

Hæc subregio est ea pars regionis Algarum olivacearum, quæ littori proxima est. Maximam partem fundi freti occupat, nonnullis locis plura milliaria a littore, velut extra Skovshoved, patens. Maris quasi savannæ sunt; nam zostera marina, quæ hinc dominans habitu tam gramineo est, ut a piscatoribus gramen marinum vocetur, fundo per spatia magna eandem uniformitatem, quæ savannarum tropicarum propria est, tribuit. Quæ uniformitas aut spatiolis nudis, ubi arena omni vegetatione caret, aut fruticetis Fuci vesiculosi et Fuci serrati intermittitur. Diversa fundi indoles efficit, ut aut species generis Fuci dominentur, aut Zostera marina; ubi arena paucis lapidibus commixta est, hæc, ut in media freti parte, ubi vero magna lapidum multitudo adest, ut in boreali et australi, Fucoideæ dominantur. Præter species generis Fuci, quæ antea nominatæ sunt, huius regionis maxime propriæ sunt hæ Melanospermeæ: *Halidrys siliquosa*, *Desmarestia aculeata*, *Dichloria viridis*, *Sporochnus rhizodes*, *Chorda Filum*, *Chordaria flagelliformis*, *Chordaria tuberculosa*, *Clavatella difformis*.

Nota 1. Munimentum Trekroner velut agger alpinus in hac regione eminet, quum lapidibus ad sat magnam profunditatem demersis velut saxum artificiosum constructum sit, ubi algæ, quæ diversis fluminibus apportantur, locum indoli suæ aptissimum inveniunt. Itaque hic singulis anni temporibus magna copia specierum novarum et rararum, præcipue Melanospermearum, reperiri potest. Hic optime cuiusque anni temporis vicissitudo algarum observari potest, quum singulis fere mensibus diversæ species obveniant. Aliæ vero per totum annum inveniuntur, velut Cruoria pellita, Zonarina Liebmanni, al. Ex aliis speciebus raris, quæ hic occurrunt, notandæ: Lichina confinis, Ilea fascia, Ilea fascia var. tenuior, Chorda lomentaria, Ectocarpus siliculosus, Elachista globosa, Chordaria tuberculosa, Mesogloia vermicularis, Helminthora multifida, Dumontia filiformis, Hormiscia penicilliformis, Rivularia nitida, Lyngbyea lutescens, Gallionella moniliformis, Agonium centrale. Ex lichenibus Verrucaria Maura.

Nota 2. Ex algis, quæ, in regione sequenti dominantes, non rara in hac occurrunt, notandæ sunt: Polyides rotundus, Furcellaria fastigiata, Rhodomela subfusca, Gigartina plicata, Chondri species.

Subregio Laminariearum.

(In tabula colore intense brunneo notata).

Hæc subregio, extrema pars algarum olivacearum, maxima ex parte in sequentem transit. Non tanta extensione, quanta prior, est, quum multis locis omnino desit. Extensio eius ex tabula I, ubi colore intense brunneo notata est, facile intelligitur. Cum ea parte freti, quæ ex indole fundi nomen regionis lapidariæ traxit, congruit. Maxima est extensione ad littus Scanicum, ubi a Kullen ad Landskronam pæne continua patet, ad Danicum vero meridiem solum versus ab Helsingora distinctos limites habet. Hæc subregio silva maris haberi potest; Laminariæ enim, 10—15 pedes altæ, erectæ velut arbores silvæ, confertæ sunt. Ita Laminaria latifolia, saccharina, digitata. Laminariarum stationes maxime australes extra Barsebæk sunt; hinc intelligitur, eas australi parti freti propiores esse, quam antea indicatum sit.[1])

Enumeratio Algarum olivacearum s. Melanospermearum.

§ 19.

Fucoideæ.

Halidrys siliquosa Lgb. — *Fucus vesiculosus* L. — *Fucus serratus* L. — *Fucus nodosus* L. *varietas tenuior* ad Kullaberg.

Lichineæ.

Lichina confinis Ag. Trekroner, Helsingora Hornem.

Laminarieæ.

Laminaria digitata Lamrx. — *Laminaria latifolia* Ag. — *Laminaria saccharina* Lamrx. — *Ilea fascia* Fr. in fl. scan (Laminaria fascia Ag.) Trekroner.[2])

Sporochnoideæ.

Desmarestia aculeata Lamrx. Kullaberg. — Taarbæk. — *Desmar. acul. var. complanata* Ag. cum præcedente. — *Desmarest. aculeata var. plumosa.* Hornem. (Ectocarpus densus Lgb.) inter Helsingburgiam et Landskronam *Dichloria viridis* Grev. Kullen, Hveen.

Dictyoteæ.

Chorda Filum Lamrx., *Chorda lomentaria* Lgb. Trekro-

---

1) I. Agardh, l. c. p. 3.

2) Laminaria Fascia var. tenuior Lgb. nullas præbet differentias specificas, ut I. Agardh opinatus est (loc. cit. p. 15).

ner primo vero 1842. *Asperococcus echinatus* Grev. — *Punctaria undulata* J. Ag. (non Laminaria Fascia var. tenuior Lgb.) in zostera marina. — *Dictyosiphon foeniculaceus* Ag. Kullen. — Trekroner. Num peculiaris est species? eam et varietates nonnullas Chordiariæ flagelliformis ego saltem distinguere non potui. — *Zonarina Liebmanni* nob.; Zonaria deusta Lgb., in palis et lapidibus ad munimentum Trekroner[1].

Ectocarpeæ.

Cladostephus verticillatus Lgb. prope Helsingoram Liebmann. — *Ectocarpus siliculosus* Lgb. Trekroner. — *Ectocarpus littoralis* Lgb. — *Ectocarpus compactus* Ag. — *Ectocarpus brachiatus* Ag. — *Ectocarpus tomentosus* Lgb. ad insulam Hveen. — *Sphacelaria cespitula* Lgb. Trekroner. — *Sphacelaria cirrhosa* Ag. ad Kullaberg. — *Sphacelaria plumosa* Lgb. Kullen Hveen. —

1) Hæc planta iam diu ab omnibus algologicis cognita est, sed mire in ea interpretanda erratum est. — Ab Lyngbyeo ad rupes maris Norvegici detecta et ad Zonariam deustam Flor. Dan. relata est. Eam ita recte definitam esse, cl. Agardhius dubitans, crustam bacilarem cuiusdam fuci habuit. Cum hanc plantam primum vidi, eam quidem Zonariæ generis affinem existimans, non facile intelligere potui, quo modo Lyngbyeus eam ad Zonariam deustam Flor. Dan. referre potuisset; quod tamen ita esse auctoritate Hofman Bangii, quæ hac in re ferme eadem ac Lyngbyei est, mihi persuasum est. Et Liebmannus ab itinere Mexicano redux, hanc sententiam speciminibus authenticis confirmavit. Distinctissimum genus, Zonariæ quidem affine, primo adspectu facile distinguitur superficie interiore tota saxis vel palis arctissime adnascente. Accuratius examinanti multi alii characteres apparent. Cum vero specimina huius plantæ in plerisque herbariis adsint, itaque facile examinari possint, hic tantum ad descriptionem et figuras, quas Liebmannus secundum specimina a me lecta in Flor. Dan. faciendas curabit, rejiciam.

Elachista fuciola Fr. — *Elachista stellaris* Areschoug Linnea V. 16 p. 233 ad Kullaberg? — *Elachista globosa* nob. ad Trekroner[1]). — *Cruoria pellita* Fr. in fl. scan., Erythroclathrus pellitus Liebm. in Kroyers Tidsskrift 2 Bind p. 189.

Chordarieæ.

Chordaria flagelliformis Ag. Kullen — Trekroner. — *Chordaria tuberculosa* Lgb. Trekroner[2]). — *Chordaria rhizodes* (Chordaria rhizodes et Ch. paradoxa Lgb.) — *Clavatella difformis* Fr. (Chorynephora marîna Ag.)

Regio Algarum purpurearum s. Rhodospermearum.

(in tabula colore purpureo notata).

§ 20.

Hæc regio ultimus vegetationis freti terminus est. Algæ purpureæ profunditatis 8—20 orgyiarum propriæ

---

1) Hæc species a ceteris huius generis facile destinguitur et forma hemisphærica vel subglobosa et eo, quod filorum maxima pars in massam densam gelatinosam coniuncta est, et ita pars eorum libera brevissima. Interdum tamen obvenit filis liberis longissimis et ita Elachistæ fucicolæ similior fit. Vide Flor. Dan. Tab. ined.

2) Hæc rara planta ferme per totum annum ad Trekroner obvenit. Nomen specificum Lyngbyei Agardhiano præferendum est, nam primum in Flor. Dan. sub nomine Ceramii tuberculosi descripta, postea a cl. Agardhio Chætophora nodulosa appellata est. Fortasse hæc et Ch. rhizodes ad proprium genus referendæ sunt.

sunt. Modo paucis freti locis hæc regio distincte circumscripta est, et hic cum extrema subregione prioris regionis, Laminariearum, pæne confluit. Algis enim purpureis sicut Laminarieis zona lapidaria maxime favet, sed ubi momenta exteriora apta adsunt, in maiorem profunditatem progrediuntur. Nullus enim locus freti tam profundus est, quin algæ purpureæ ibi crescere possint, dummodo fundum sibi aptum vel lapidum vel testarum habeant; quum autem maxima fundi pars extra zonam lapidariam argilla mollissima tecta sit, raro hic obveniunt. Recte cl. Agardhius iam indicavit, Rhodospermearum et Chlorospermearum stationes sibi retrorsum respondere, quum illæ in septentrionali freti parte, hæ in australi maxime dominentur. Sed algæ purpureæ australi parti freti plus, quam ille præstantissimus algologus putat, appropinquant; nam septentrionem versus ab Hveen multæ reperiuntur Florideæ, et etiam ad Stevns tres species Dellesseriarum aliæque, magnopere quidem contractæ, obveniunt. Huius regionis maxime propriæ species sunt: Iridea edulis, Delesseriæ, Hutchinsiæ, Callithamnii, Ceramii, Gigartinæ species, Odonthalia dendata.

Nota. Ex iis, quæ allata sunt, facile intelligitur, in diversis regionibus non eas Algas solas reperiri, a quibus nomen ductum sit; in omnibus enim regionibus Algæ omnium sectionum obveniunt, sed in regione sua quæque sectio tot species proprias continet, ut ceteræ præ iis pæne evanescere videantur. Nomina igitur ex plantis cuiusque regionis maxime propriis tracta sunt.

§ 21.

Enumeratio Algarum purpurearum s. Rhodospermearum.

Gloiocladeæ.

Mesogloia vermicularis Ag. Kullen — Trekroner. — *Ægira Zosteræ* Fr. (Mesogloiia Zosteræ Areschoug Alg. scandin. exsic. fasc. III 67) Helsingora Liebm. Havnia. — *Helminthora multifida* Fr. (Chordaria multifida Lgb.) Kullen-Trekroner.

Gastrocarpeæ.

Dumontia filiformis Grev. (Gastridium filiforme Lgb.) Hveen-Trekroner. — *Iridea edulis* Bory (Halymenia edulis Ag.) Kullen-Helsingb.

Spongiocarpeæ.

Polyides rotundus Grev. Helsing. —

Furcellarieæ.

Furcellaria fastigiata Lmrx. —

Florideæ.

Delesseria sanguinea Lmrx. — *Delesseria sinuosa* Lmrx. — *Delesseria alata* Lmrx. — *Rhodomenia palmata* Grev. — *Odonthalia dentata* Lgb. Kullen-Hellebæk. — *Rhodomela subfusca* Ag. — Gigartina purpurascens Lmrx. Kullen-Hveen. — *Gigartina confervoides* Lmrx. cum præcedente. — *Gigartina plicata* Lmrx. — *Gigartina plicata var. hippuroides* (Scytosiphon hippuroides Lgb.) ad Hellebæk. — *Chondrus crispus* Lgb. — *Chondrus crispus var incurvatus* ad Hellebæk. — *Chondrus membranifolius* Grev. — *Chondrus Brodiæi.* Grev. — *Chon. Brodiæi angustissimus* Suhr ad Hellebæk. — *Chondrus Brodiæi ligulatus* Suhr ad Hellebæk. — *Sphærococcus rubens* Ag. ad Hellebæk. — *Ptiota plumosa* ad Helsingoram Liebm. ad Kullaberg.

Ceramieæ.

Hutchinsia Brodiæi Ag. ad Helsingoram Liebm. — *Hutchinsia penicillata* Ag. ibidem. — *Hutchinsia byssoides* Grev. ad Kullaberg. — *Hutchinsia tenuis* Ag. ad Lomna Agardh. — *Hutchinsia nigrescens* Ag. — *Hutchinsia violacea* Lgb. — *Hutchinsia roseola* Ag. (H. stricta Ag. in Syn.) Hellebæk, Kullaberg. — *Hutchinsia divaricata* ad Helsing. Liebm. — *Hutchinsia lepadicola* Lgb. ad Helsing. Liebm. — *Ceramium rubrum* Ag. — *Ceramium diaphanum* Roth. — *Ceramium ciliatum* Ducluz. — *Ceramium elongatum* Roth. Kullen-Landskrona. — *Ceramium elong. var denudatum* Ag, (Ceramium brachygonium Lgb.) et *Ceram. elongatum proliferum* Ag. cum præcedente. — *Callithamnium pyramidatum* Liebm. Helsing. Liebm. — *Callithamnium Rothii* ad Kullaberg. I. Agardh. — *Callithamnium fruticulosum* Lgb. ad Kullaberg. — *Callithamnium lanuginosum* Lgb. in Ceram. rubro Fries. — *Callithamnium repens* Lgb. Fries in fl. scan. — *Callithamnium thyoides* Ag. ad Kullaberg Ag. Fr.

§ 21.

Extensionem algarum horizontalem contemplantes facile perspicimus, eam in directione sola longitudinali freti sive a septentrione ad meridiem adesse, non vero ab oriente ad occasum, quum hac ratione eadem sit, ac extensio verticalis, quia maxima freti pars ex duobus planis proclivibus, quæ convallem inter Sjællandiam et Scaniam constituunt, formatur, et pars, quæ intermittitur, ut iam antea demonstravimus, omni vegetatione caret. Quomodo vegetatio in directione ab meridie septentrionem versus mari salso et minus salso diversa ratione afficiatur, occasionem optimam observandi præbet fretum; fretum enim Öresundicum, ut antea vidimus, unum ex iis est, in quibus maris Codani salsa et Baltici pæne dulcis aqua commiscetur. Itaque in boreali freti parte algæ purpureæ, in media olivaceæ, in australi virides dominantur. Quæ distributionis diversitas in tabula I. coloribus respondentibus indicata est. Algæ vero purpureæ quasi coactæ principatum maris abdicant; in contractissimas enim formas potius abeunt, quam omnino evanescant. Formæ enim generum Delesseriæ, Chondri aliorumque, quæ etiam ad Stevns obveniunt, parva sunt magnitudine et colore pæne viridi, ut initio dubium sit, num eædem species sint.

De momentis, quibus distributio Algarum secundum regiones constituitur.

§ 23.

Modus, quo algæ vulgo collectæ sunt, leges generales distributionis geographicæ proferre non potuit, quum omnes species, in littus congestæ, itaque a locis primariis divulsæ, colligerentur. I. Agardh primus in libro, quem sæpe citavi, exemplum optimum præbuit. Momenta, quæ distributionem algarum constituunt hæc affert: 1) chemicam maris indolem, 2) altitudinem maris, 3) tranquillitatem maris et violentiam diversam, 4) qualitatem soli matricis. Quum algæ duo diversa media, fundum et mare circumdans, attingant, momenta, quibus afficiuntur, hoc modo dividenda sunt:

1. Fundi { a) chemica indoles.
   b) mechanica indoles.

2. Maris { a) chemica indoles.
   b) profunditas. { aa) indoles lucis.
   bb) indoles motus.

---

§ 24.

Chemicam indolem fundi nullius esse momenti ad vegetationem algarum constituendam, facile intelligitur, quum radices algarum tantummodo affigant, nec humores nutrientes recipiant. Chemica igitur indoles fundi, mari circumdante soluti, modo indirecte afficere potest.

Indoles autem mechanica fundi ad stationes algarum constituendas, secundum diversitatem aggregationis, maximi momenti est. Algæ enim radices nonnisi rebus solidis affigere possunt; unde hæc lex evadit, ut Algæ omnibus locis freti, in quibus fundus lapidibus careat, desint. Quum vero minor pars fundi ex saxis vel lapidibus congestis, maior vero ex arena aut argilla constet, hinc, perspicuum est, minorem fundi partem Algis obductam esse, itemque earum propria domicilia esse, ubi maxima lapidum multitudo est; quod ita quoque in zona lapidaria esse, ubi Algæ et purpureæ et olivaceæ summam vegetationis massam congesserint, iam vidimus. Zostera autem marina, quæ ad Algas non pertinens nonnisi in arena soluta radices figere potest, propriam zonæ arenariæ vegetationem constituit. Quæ quum maximam freti partem occupet, Zostera marina omnes Algas massa vegetationis superat. In argilla mollissima neque Zostera marina neque algæ radices figere possunt, itaque zona tota argillacea omni vegetatione caret. Et indoles maris chemica ad faciem vegetationis suæ determinandam plurimum valet; quod diversitas vegetationis borealis et meridionalis freti partis satis declarat[1]). Diversa salsitudo partium freti efficit, ut, algis purpureis in australi parte pæne evanescentibus, virides dominentur.

Secundum chemicam aquæ indolem vix quidquam tanti momenti algis est, quanti discrimen lucis, profunditate di-

---

[1]) v. J. Agardh. l. c p. 4.

versa effectum. Quæ lucis vis nondum satis accurate examinata est, quum sola intensitas lucis diversitate profunditatis mutari dicta sit[1]); sed aliud quiddam fieri, iam videbimus. Ars enim optica demonstrat, radios varii coloris, ex quibus lux decolor composita sit, in humorem non omnino pellucidum (qua indole mare est) penetrantes, diverso modo frangi, ut non omnes æque profunde per mare penetrent. Rubri radii ad maximam profunditatem perveniunt, deinde aurantiaci, deinceps ceteri iusto ordine, ut igitur cærulei et violacei minime intrent. Quæ lucis indoles ex aurora matutina et vespertina, quum soli radii rubri et flavi per aerem vaporibus plenum penetrent, nobis nota est. Hinc intelligitur, ut præterea urinatorum usus docet, in maiorem maris profunditatem radios rubros pæne solos pervenire. Sed in superioribus vidimus, sectiones algarum, quæ diversi coloris sint, in diversa profunditate crescere; itaque algas secundum eum ordinem, quo a superficie maris ad profunditatem proveniunt, itemque radios secundum ordinem, quo profundius penetrant, enumerantes, videmus, sectiones algarum, quæ diversi coloris sint, iis locis obvenire, quibus maximam abundantiam ejusdem coloris radiorum esse supra demonstravimus.

Color.	*Algæ.*	*Profunditas.*
Radii violacei — cyanei — caerulei	Algæ viridicaerulescentes (Oscillatorineæ)	Superficies.
— virides	Algæ virides (Chlorospermeæ)	Ped. 10—25.
— flavi — aurantiaci	Algæ olivaceæ (Melanospermeæ)	Ped. 25—50.
— rubri	Algæ purpureæ (Rhodospermeæ)	Ped. 50—65.

Nequaquam nobis sumimus, ut phænomena difficillima, quæ varii algarum colores præbent, omnino ex-

[1]) v. J. Agardh. l. c.

pediverimus; tum enim nobis explicandum erat, cur algæ eiusdem sint coloris, ac radii lucis circumdantis; sed putamus, nos, his præmissis, ad explicationem propius accessisse.

Præter vim, quam habet profunditas ad indolem lucis constituendam, eadem quoque ad effectum undarum in fundum statuendum maximi momenti est, quum effectus earum profunditati retrorsum respondeat. Ex quo hæc lex generalis evadit, ut algæ virides crescant, ubi summa violentia undarum est, quum minima in profunditate obveniant, deinde olivaceæ, denique purpureæ, ubi violentia minima. Sed præter motum maris, qui, undis effectus, maiorem profunditatem non attingit, alius est, qui, flumine effectus, sæpe maximus est, ubi motus undarum minimus est. Hoc ita in freto se habet, ubi Laminarieæ et Rhodospermeæ reperiuntur. Hic profunditas tanta est, ut motus undarum ad fundum pervenire nequeat; sed hic quoque vis fluminis maxima est. Ex quo intelligitur, Algas, quibus flumen faveat, in mari, violentiæ undarum exposito, evanescere, et contra. Notandum est, motum fluminis etiam indirecte vegetationi algarum favere, quum indolem fundi, huic aptissimam efficiat; itaque difficile est constituere, utrum indoles fundi an flumen vegetationem Algarum plus adiuvet. Hoc schema conspectum momentorum, quæ vegetationem algarum constituunt, præbet.

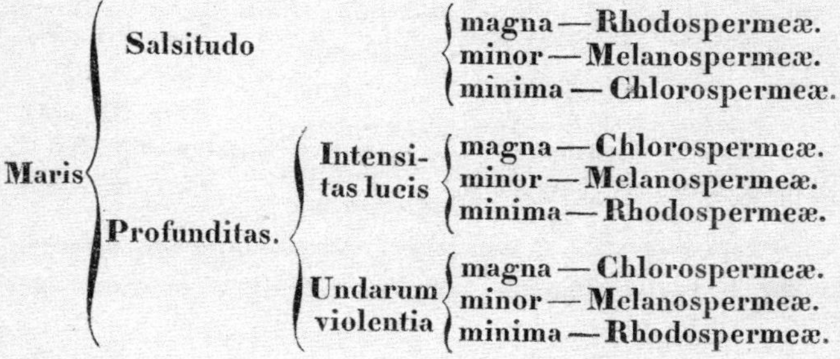

Maris
Salsitudo
magna — Rhodospermeæ.
minor — Melanospermeæ.
minima — Chlorospermeæ.
Profunditas.
Intensitas lucis
magna — Chlorospermeæ.
minor — Melanospermeæ.
minima — Rhodospermeæ.
Undarum violentia
magna — Chlorospermeæ.
minor — Melanospermeæ.
minima — Rhodospermeæ.

§ 25.

Agardhius primus probavit, Algis proprium esse, notas gravissimas, quæ characteres subordinum constituant, ex coloribus arcessendas esse[1]). Hoc divisionis fundamentum ab omnibus hodie adoptatum est, ut vix quisquam dubitet, quin naturæ maxime congruat divisio Algarum in *virides, olivaceas, purpureas.* Nunquam vero explicatum est, cur Algæ hac in re ab aliis plantis tantum recedant, cur color, qui alias nota specierum sit, hic character gravissimi momenti sit. Supra vidimus, Algas, quæ diversi coloris sint, in diversa maris profunditate sive in diversis regionibus crescentes non eodem modo luce affici; qua quidem re sola hæc proprietas summa expediri potest. Algæ enim sunt unicus ordo, cuius diversi subordines diverso modo luce afficiuntur, ut non mirandum sit eas, proprietate coloris a ceteris plantis recedere. In stationibus igitur propriis harum plantarum inest causa, cur color ad eas systematice distribuendas plurimum valeat. Quam distributionem cum natura congruere, ex hoc quoque perspicuun est, quod ex membris analogis illi tres subordines compositi sunt; hic enim ut semper in omni naturali systemate evolutionem per series parallelas cum membris analogicis invenimus; quod divisionis fundamentum vix usquam in systemate tanta constantia, quanta ab Agardhio in Algis distribuendis adhibitum est[2]). Conspectum, in his principiis positum, familiarum Algarum proponimus:

---

1) v. Systema algarum p. XII, et Species algarum p. LXXII.
2) Spec. Algar. p. LXXII.

Ordo Algæ.

1. *Subordo* Melanospermeæ.
   1. *Familia.* Fucoideæ[1]).
   2. *Familia.* Laminarieæ.
   3. *Familia.* Ectocarpeæ.

2. *Subordo* Rhodospermeæ.
   1. *Familia.* Florideæ.
   2. *Familia.* Halymenieæ.
   3. *Familia.* Ceramieæ.

3. *Subordo* Chlorospermeæ.
   1. *Familia.* Ulvaceæ.
   2. *Familia.* Corfervoideæ.
   3. *Familia.* Diatomaceæ.

In duobus prioribus subordinibus (Melanospermeis et Rhodospermeis) priores familiæ (Fucoideæ et Florideæ) extensionem æquabilem per omnes dimensiones significant et corporibus respondent; secundæ (Laminarieæ et Halymenieæ) extensionem prægnantem per duas dimensiones indicant, planis respondentes, tertiæ (Ectocarpeæ et Ceramieæ) extensionem prægnantem per unam, lineis respondentes. Subordini ultimo membrum, familiæ perfectissimæ subordinum superiorum analogum, deest; sed novum, Diotamaceæ, accedit, quod inferius est, quam cujus analogum membrum in superioribus sit. In ultimo igitur subordine nullum est tam perfectum membrum, quod corporibus respondeat, sed non nisi planis et lineis (Ulvaceis et Confervoideis), et altera ex parte aliud novum accedit, quod tam imperfectum est, ut analogo in superioribus omnino careat, quod puncto respondet (Diatomaceæ). Diatomaceas enim ad Chlorospermeas referendas esse puto, quamquam pleræque colore cum Melanospermeis congruunt; nullum enim dubium est, quin inter algas structura infima sint.

---

1) Sporochonoideis, Dictyoteis, et Chordarieis incl.

Nota. Numerus specierum omnium plantarum, quæ in freto obveniunt, centum septuaginta trium est. Nam præter centum sexaginta sex species ex algarum ordine hæ plantæ hic reperiuntur: ex Naiadeis: Ruppia rostellata Koch, Ruppia maritima L., Zanichellia palustris L. Zostera marina L., ex Characeis: Chara Baltica Fr. et Aspg., Chara nidifica [1]) Engl. Bot., ex Licheneis Verrucaria Maura Fr. Species inter diversas sectiones algarum ita distributæ sunt:

Melanospermeæ	36
Rhodospermeæ	41
Chlorospermeæ	51
Diatomaceæ	38
	166.

1) Hanc plantam pæne ignotam profunditate 8 - 10 orgyiarum inter Taarbæk et Hveen inveni. Eam distinctissimum genus, quodammodo transitum inter Characeas et Chlorospermeas (præcipue Siphoneas) efficiens, constituere omni dubio vacare puto.

CAPUT TERTIUM.

De regionibus animalium in freto.

§ 26.

In præmonendis iam notavi, quid regio intelligatur ideoque statim ad ea examinanda, quæ adhuc in hac ratione geographiæ animalium marinorum scriptis tradita sunt, aggrediar. Audouin et Edwards primi, littora francogallica perscrutantes, diversitatem magnam regni animalium per diversas altitudines partis littorum æstu nudatæ animadverterunt; quam secundum diversitatem hæc in quatuor regiones diviserunt[1]). Eædem rationes a Sarsio ad littora Bergensia repertæ sunt[2]). Regiones hæ sunt:

1. *Regio Balanorum* partem superiorem proximam terræ occupat. Præter Balanos hic præcipue Purpura lapillus dominatur.

2. *Regio Patellarum*, quæ inferior est, huic Patellæ, Turbines, Mytilus edulis, Actiniæ propriæ sunt; ex plantis Fuci.

3. *Regio Corallinarum*. Hic inprimis Corallina officinalis, Mytilus Modiolus, Ascidiæ, Spongia, Alcyonia, Annulata, et ex plantis Zostera marina, inveniuntur.

4. *Regio Laminariearum*. Nudibranchiata, Asteridæ, Caprellidæ, Pycnogonidæ dominantur.

Distributio vero animalium in maiore profunditate a neutro illorum auctorum examinata est. Nobis indolem

---

1) Annales d. scienc natur. Vol. 21 p. 26

2) Beskrivelser og Iagttagelser over nogle mærkelige eller nye i Havet ved den bergenske Kyst levende Dyr p. VI 1835.

littorum Francogalliæ et Norvegiæ, adhuc examinatorum, cum iis, quæ nunc examini subiiciemus, comparantibus, diversitas summa distributionis animalium propter horum diversissimam naturam confestim expectanda erit. Illa littora sunt prærupta saxosaque, quorum magna spatia æstu aut teguntur aqua aut nudantur; hæc vero planissima effectui maris alternatim crescentis et decrescentis non exposita sunt. Itaque facile intelligimus, quo planiora littora sint, eo longiores singulas regiones esse; hic enim nonnisi per maius spatium eadem diversitas altitudinis maris, quæ ad littus præruptum per multo minus spatium, apparet. Itaque infra videbimus, si regiones quatuor, quas commemoravimus, per maius spatium quasi extensas animo fingemus, eas cum iis, quas in freto reperiamus, magna ex parte congruere.

---

§ 27.

Directa contemplatione distributionem animalium per regiones in freto explorari non posse, ut ad littora prærupta, quorum magna pars æstu nudetur, sed pæne ubique instrumento peculiari animalia, fundum habitantia, in lucem proferenda esse, iam antea notavi. Hoc modo fundum freti per totam longitudinem per transversum a Danico ad Scanicum littus perscrutans, in profunditate diversa species diversas inveniri, quin etiam tota genera familiasque profunditatum diversarum propria esse, comperi. Quas secundum diversitates fundus in diversas regiones dividi potest; quarum nomina ex animalibus earum propriis arcessenda sunt, inprimis ex iis, quorum divisiones superiores (i. e. ordines et familiæ) diversitati locorum respondent; ergo ex iis animalibus, quorum non modo diversæ species et genera, sed etiam familiæ vel ordines una cum gravi momentorum externorum diversitate apparent. Id saltem quod ad fretum attinet, præcipue in Mollusca convenire puto; quamobrem postea nomina ex iis arcessivi. Ex tribus regionibus, quæ infra explicabuntur, *regio Trochoideorum* littori adiacens totam zonam arenarium occupat; extra hanc *regio Gymnobranchiorum* cum zona lapidaria et Laminariearum Algarumque purpurearum congruit; denique *regio Buccinoineorum*, quæ zonam argillaceam occupans regio quoque profunditatis nominari potest.

---

§ 28.

Regio Trochoideorum.

(In tabula colore coeruleo, viridi, dilute bruneo et albo notata).

Hæc regio a littore et æque longe ac fundus arenarius patet i. e. a 0 ad 7—8 orgyias. Kullen unicus locus est, ubi pæne omnino deest, quia profunditas littus attingit. Indoles huius regionis generalis iam ex antecedentibus magna ex parte perspicua est. Ex quibus cognitum est, hic undarum violentiam maximam esse. Motus undarum ubique in hac zona fundum ergo etiam animalia in eo habitantia attingere possunt; sed effectus in fundum profunditati retrorsum respondet. Omnia igitur animalia, quæ hic degunt, aliquo modo se a vi undarum perniciosa defendere possint, necesse est. Quod duobus modis fieri videbimus. In animalibus, quæ testa calcarea prædita sunt, hæc crassitudinem et duritiem magnam assequitur; cuius rei documento inprimis Littorina littorea est, quæ sæpe inter lapides littoris undis sine ullo detrimento agitur Animalia autem testis destituta alio modo defendenda natura curavit; in cuniculos enim, in arena fossos, se abdere possunt, ut non solum motum undarum, sed etiam vim aëris, quo, undis recedentibus, afficerentur, evitent. Huius enim regionis proprium est, quod magna eius pars omnino aqua, mari recedente, nudatur. Quamquam enim in boreali sola parte æstuum accessus et recessus apparet, tamen altitudo maris secundum flumen diversum alternat. Ex animalibus, quæ sic se abdunt, nominanda: *Mya arenaria*, *Nereis diversicolor*, *Arenicola piscatorum*, *Corophium longicorne* [1]). Non exigui momenti animalibus est vegetatio copiosa, quæ huius zonæ propria est, et magna eius varietas. Ut Ulvæ suas proprias formas alunt (velut Paludinella Ulvæ, Planaria Ulvæ), sic fruticeta densa Fucoidearum suas formas habent (velut Heteronereis fucicola), et savannæ late patentes-Zosteræ marinæ aliis favent (v. Terebella zostericola, Nereis zostericola).

Huius regionis regnum animalium paucis his verbis describi potest: *animalia pleraque hic phytophaga aut dura testa aut facultate se in arenam abdendi contra vim perniciosam undarum et aëris armata.*

Propriæ species, quæ pæne ubique in hac regione reperiuntur, nominandæ sunt: *Crangon vulgaris* et *Mysis flexuosus*, quæ littori quam maxime appropinquant. *Palæmon Squilla* simul cum *Idotea tricuspidata* in paulo maiore profunditate gregatim obvenit; in locis tranquillioribus *Jæra albifrons* et *Corophium longicorne* dominantur. *Talitrus saltator* et *Orchestia littorea* inter Fucoideas congestas latent. *Balanus balanoides* lapides in ipso limite maris obducit. *Nereis diversicolor* cuniculos in arena præci-

---

1) Vide: Krøyers naturhist. Tidsskr 3 Bd. p. 558. Beretning om en Excursion o. s. v. af Ørsted.

pue sub lapidibus semper quam maxime socialiter habitat. *Spionis seticornis* tubi arenarei in littore plano magna multitudine verticaliter positi obveniunt. *Arenicolæ piscatorum* immensus numerus excrementis spiraliter tortis ad aperturam cuniculi indicatur. *Lumbricillus linearis* et *Gordius marinus* glomerulos formantes sub unoquoque littorum lapide apparent. *Littorina littorea, retusa et fabalis* saxa, plerumque aqua nudata, potius quam mare habitant. *Paludinella Ulvæ et vulgaris, Mytilus edulis, Cardium edule, Tellina Baltica et Mya arenaria* in paulo maiore profunditate obveniunt. *Echinus esculentus* non nisi in ultima huius regionis parte ɔ: in profunditate 5—8 orgyiarum simul cum *Asteracanthio violaceo* reperitur. *Campanularia geniculata* et *Flustra membranacea* plantas et palos obducunt.

Nota 1. Hæc regio in complures subregiones dividi quidem potest, quibus tamen, arctissime inter se coniunctis, non totæ familiæ ut regionibus sed modo genera vel species propria sunt. Quamquam ita Littorinæ sæpe in extrema regione obveniunt, tamen maxime ad partem littori proximam commorantur, ut pæne amphibiæ haberi possint, quum æque in aqua ac super eam vivant. Itaque fortasse hæ subregiones constitui possunt.

1. *Subregio Littorinarum* spatium littoris, quod, aqua recedente denudatur, occupat. Præter Littorinas species generum Nereidis, Spionis, Arenicolæ, Myæ, quæ supra nominatæ sunt, huius subregionis propriæ sunt.

2. *Subregio Mytili edulis* ab extremo prioris subregionis termino usque ad ea loca, quibus arena cum argilla commixta est, patet. Mytilus edulis, quamquam in superiore subregione quoque reperitur, hic tamen dominatur. Maxime propriæ species sunt: Akera bullata, Ascidia intestinalis, Carcinus Moenas, Terebella zostericola, Echinus esculentus, Lucernaria quadricornis. Ex plantis Zostera marina hic dominatur.

3. *Subregio Nassæ reticulatæ* extremam huius regionis partem, ubi arena iam cum argilla commixta et vegetatione destituta est, occupat. Hæc subregio tantum invenitur, ubi regio lapidaria

deest, quæ si adest, distinctiorem limitem inter zonam arenariam et argillaceam efficit. Sin deest, ut pæne in toto medio freto, species huius regionis et profunditatis inter se commixtæ reperiuntur. Duas tantum species huius regionis proprias: Nassam reticulatam et Corbulam nucleum, nominare possum.

Nota 2. In hac regione peculiaris fundus cum propriis animalium speciebus aliquot locis invenitur, *limus*, qui idem in mari atque humus in terra est; nam ex plantis putredine solutis oritur. Massam atram mollissimam et pinguissimam constituit, quæ præcipue Ulvæ lactucæ et latissimæ favet. Oritur in locis depressioribus, quibus plantæ congestæ dissolvuntur, nec flumine aqua renovatur. Propriæ species nominandæ sunt: Bullæa (aperta?), Amphitrite auricoma, Lumbricillus verrucosus, Nemertes olivacea et melanocephala

Nota 3. Quamquam pisces minime omnium animalium marinorum certis locis obnoxii sunt, quum libere circumnatent, nonnulli tamen huius regionis proprii sunt. Ex quibus imprimis nominandi: *Gasterosteus aculeatus*, parvulus quidem, sed fortis et validus piscis, qui prope littus circumnatans certamina mira init; *Spinachia vulgaris*, quæ turbas Palæmonis Squillæ persequens ova, quæ ab his paruntur, comedit; *Cottus scorpius*, qui tardus, stupidus voraxque inter folia Zosteræ marinæ ad fundum se lente movet; *Gunellus vulgaris*, qui inter lapides et fucos quiete degens interdum, aqua recedente, in arido relinquitur; *Zoarces viviparus* qui item inter lapides et fucos, ventre magna multitudine pullorum tumente, morans, Mytilo eduli præcipue alitur; præterea *Platessa Flesus*, interdum fluvios intrans, *Ammodytes Tobianus*, *Angvilla vulgaris* [1]).

Nota 4. Hac regione cum regionibus littorum Francogallicorum et Norvegicorum comparata, facile intelligitur, eam cum tribus prioribus regionibus, Balanorum, Patellarum, Corallinarum congruere. Hoc vero discrimen præcipue apparet, quod Patellæ et Corallinæ in freto regionis sequentis propriæ habendæ sunt.

---

1) v. Krøyer Danmarks Fiske. Hæft. 1—3.

§ 29.

Enumeratio specierum omnium animalium in regione Trochoideorum.

Crustacea.

Geryon tridens Kr. Gilleleie, Lyngbye. — *Carcinus Moenas* Leach. — *Pinnotheres Pisum* Latr. Gilleleie Lyngbye. — *Lithodes arctica* Latr. in boreali freti parte Kröyer. — *Crangon vulgaris* Fabr. — *Hippolyte Gaimardii* Edw. Hellebæk. — *Palemon Squilla* Fabr. — *Mysis flexuosus* Lamk. — *Talitrus saltator* Edw. — *Orchestia littorea* Leach. — *Metoecus Medusarum* Kr. in *Medusa aurita.* — *Hyperia sp. nov?* cum præcedente in Med. aur. — *Gammarus Sabbini* Leach. Hellebæk — Hveen. — *Gammarus Locusta* Fabr. — *Idotea pelagica* Leach. — *Idotea tricuspidata* Desm. — *Idotea emarginata* Fabr.[1]) — *Anthura* (*arctica* Kr.?) Kallebodstrand. — *Tanais Örstedii* Kr. ibidem. — *Tanais Curculio* Kr. cum præcedente. — *Jæra albifrons* Leach. — *Sphæroma* sp. nov.?

Arachnida.

Trachheariæ. Acaridæ.

Acarus Basteri Johnst[2]) in Kallebodstrand. — *Acarus setosus* nob.[3]) ad Trekroner.

---

1) Jam pridem dubitationem meam de harum trium specierum diversitate attuli (vide Kröyers Tidssk. 3 B. p. 561.)

2) Loudons Mag. of nat. hist. V. 9, p. 353.

3) Corpore cinereo oblongo-ovali, et anticam et posticam partem versus constricto, postice brevissime acuminato, palpis sub rostro obtuso absconditis, mandibulis? pedibus remigatoriis æqualibus, setis longissimis numerosis in postico, modo duobus in antico corpore; long. $\frac{1}{3}'''$
Utraque species distinctissima genera constituit

Annulata [1]).

Lepidonote impar Örsd. Taarbæk. — *Lepidonote cirrata* Örsd. — *Pholoe Baltica* Örsd. — *Heteronereis fucicola* Örsd. Hellebæk. — *Nereilepas variabilis* Örsd. Hellebæk. — *Nereis zostericola* Örsd. Hellebæk. — *Nereis diversicolor* Mül. — *Phyllodoce assimilis* Örsd. Kullen. — *Phyllodoce mucosa* Örsd. — *Leucodorum ciliatum* Johnst inter Havniam et Trekroner. — *Spio seticornis* O. Fabr. ad Ny Badehuus. — *Spio filicornis* O. Fabr. Helsingborg. — *Arenicola piscatorum* Lamk. — *Spirorbis nautiloides* Lamk. — *Amphicora Sabella* Ehrb. (*Tubularia Fabricia* Faun. groenl. fig. 12, Nais equisetina Düges. Ann. d. sc. nat. Tom. VIII. Pl. I. f. 24). — *Terebella zostericola* nob. Issefjord. — *Amphictene auricoma* Sav. — *Tubifex serpentinus* nob. prope Taarbæk [2]). — *Lumbricillus verrucosus* nob. Kallebodstrand. — *Lumbricillus lineatus* Örsd. [3]) ibidem. — *Nais littoralis* Örsd. (Kröyers Tidsskrift 3 B. p. 136.) Kallebodstrand. — *Nais elinguis* Müll. cum præcedente.

Apoda [4]).

Dendrocoelum lacteum Örsd. Havnia — Stevns. — *Planaria torva* Mül. cum præcedente. — *Planaria Ulvæ*

---

1) Pleræque species ex Annulatorum ordine, quæ ic enumerantur, descriptæ sunt in libro meo: Conspectus Annulatorum Danicorum 1843.

2) Tubificis genus differt a Lumbricillo præcipue setis dorsalibus partim capillaribus partim uncinatis.

3) Lumbricillus, Lumbricillorum familiæ typus, distinguitur setis et superioribus et inferioribus subulatis abbreviatis ferme rectis (v. Koröyers Tidsskr. 3 B. p. 130—31).

4) Omnes species ex Planarieorum ordine, quæ in sequentibus enumerantur, descriptæ sunt in libro meo: Entwurf einer systematischen Eintheilung und specieller Beschreibung der Plattwürmer 1844.

Örsd. — *Planaria affinis* Örsd. Havnia. — *Monocelis unipunctata Örsd.* — *Monocelis lineata* Örsd. — *Monocelis rutilans* Ehrb. Kallebodstrand. — *Telostoma Mytili* Örsd. — *Prostoma croceum* Örsd. Kallebodstrand. — *Prostoma suboviforme* Örsd. ibidem. — *Vortex littoralis* Örsd. ibidem. — *Vortex capitata* Örsd. Taarbæk. — *Typhloplana marina* Örsd. Hveen. — *Convoluta paradoxa* Örsd. Kallebodstrand. — *Cephalothrix coeca* Örsd. inter Havniam et Trekroner. — *Tetrastemma subpellucidum* Örsd. Snedkersteen. — *Tetrastemma bioculatum* Örsd. inter Havniam et Trekroner. — *Tetrastemma assimile* Örsd. Kallebodstrand. — *Nemertes melanocephala* Johnston Kallebodstrand. — *Memertes bioculata* Örsd. inter Havniam et Trekroner. — *Nemertes badia* Örsd. — *Nemertes olivacea* Johnst. — *Polystemma pulchrum* Örsd. inter Havniam et Trekroner. — *Gordius littoreus* Mül. — *Angvillula marina*, Vibrio marina Müll., vix Enchelidium marinum Ehrb.[1])

Mollusca.

Gasteropoda.

Limnæa Baltica Niels. Kallebodstrand. — *Paludinella Ulvæ* Beck. — *Paludinella Baltica* (Paludina Baltica Niels). — *Paludinella vulgaris* nob.[2]) — *Neritina Baltica* Beck

---

1) Act. Acad. Berol. 1835 p. 219. Cum Vibrio marina Mül. oculis omnino destituta sit, cum Enchelidio marino Ehrb. identica, ut cl. Ehrenbergius putat, esse non potest.

2) Hæ duæ species, quæ semper inter se commixtæ obveniunt, facile distinguuntur hoc modo:

Paludinella Baltica, testa oblongo-conica subperforata sordide virescente opaca vix pellucida, anfractibus subsenis convexiusculis, spira producta non acuta, pede oblongo, antice exciso postice rotundato non angustiore.

Paludinella vulgaris, testa oblongo-conica non perforata

Kallebodstrand. An varietas Neritinæ fluviatilis? — *Trochus cinerarius* Mül. Kullen — Landskrona. — *Littorina littorea* Fer. — *Littorina (Helix Mtg.) petræa* Kullen. — *Littorina* (Turbo Matton) *rudis* ibidem. — *Littorina retusa* Fer. — *Littorina (Turbo) fabalis Turt.* — *Lacuna quadrifasciata* Turt. — *Lacuna canalis* Turt. — *Lacuna pallidula* Turt. Trekroner. — *Nassa reticulata* Aut. — *Purpura Lapillus* Lamk. Kullen — Helsingborg. — *Cerithium Danicum* Beck. Issefjord. — *Akera bullata* Mül. — *Bullæaa perta* Lamk.? Trekroner. — *Tergipes lacinulatus* Cuv. Helsingora — Havnia. — *Limapontia nigra* Johnst[1]) (Planaria limacina O. Fabr., Planaria capitata Müll.).

Acephala.

Mytilus edulis L. — *Cardium edule* L. — *Corbula nucleus* Turt. — *Mactra solida* L. — *Mactra* nov. spec? Kullen. — *Tellina tenuis* Lamk. — *Tellina Baltica* L. — *Mya arenaria* L.

Tunicata.

Phalusia intestinalis Sav. — *Ascidia (Phalusia) tubifera* nob.[2])

Echinodermata.

Echinus esculentus L. — *Asteracanthion violaceus* M. L. (Asterias rubra et A. violacea Mül.).

---

pellucida nitida fusco-lineata, anfractibus subsenis minus convexis, spira producta acuta, pede oblongo antice rotundato postice sensim angustiore.

Præterea quatuor species, quæ ad hoc genus pertinere videntur, ad Kullaberg inveni; quod tamen decidere non audeo, quia omnia specimina animalibus destituta erant.

1) Loud. Mag. of. nat. hist. V. p 9 79.

2) Corpore globoso gelatinoso subpellucido griseo-albescente, aperturis in tubos productis, quorum altero æque longo ac toto corpore, altero duplo vel triplo breviore; apertura huius excisuris 4 (?), illius 6 (?) instructa. long. 5.'''

Polypi.

Lucernaria quadricornis L. — *Coryne squamata* Lamk. in ipso limite maris. — *Lobularia digitata* Lamk. — *Campanularia geniculata* Flem. — *Flustra membranacea* L.

---

§ 30.

Regio Gymnobranchiorum.

(In tabula colore intense brunneo et purpureo notata).

Hæc regio modo parvam extensionem in freto habet, hic tamen plantarum et animalium maxima congesta est multitudo. Termini eius cum regione, quæ ex indole fundi lapidarea vel testacea et ex vegetatione Laminariearum et Rhodospermearum vocata est, congruunt. Itaque in australi freti parte omnino deest, et in boreali dominatur. Vegetatio igitur copiosa et speciosa eam ornat, et hujus regionis præterea proprium est, quod aqua continuo flumine renovatur. Itaque indole momentorum externorum ab aliis regionibus magnopere differt. Fundus quasi stratus efficit, ut animalia eum penetrare non possint; sed non tantum iis opus est, quantum in regione priore; hic enim neque motus undarum neque effectus aëris timendi sunt. Inter folia magna Laminariearum secura vivere possunt. Itaqua non mirandum est, quod animalia tantum ab animalibus prioris regionis recedunt. Iam antea exposui, hujus regionis vegetationem silvas terrarum quasi imitari; si animalia quoque respicias, silvas etiam tropicas refert. Hic non ea uniformitas apparet, quæ in savannis Zosteræ marinæ spatiosis, sed summa multitudo, copia varietasque animalium et plantarum. Hic, ut in illis silvis, fundus plantis tam dense obsitus est, ut pleraque animalia in his commorentur. *Caprellidæ*, *Pycnogonidæ*, *Gymnobranchia*, *Patellæ*, *Chitones*, *Ascidiæ*, *Holothuriæ*, *Actiniæ* cum plantis, inter quas versantur, colorum splendore et varietate certant. *Actinia candida*, *plumosa*, *coccinea*, *holsatica*, al. una cum *Holothuria phantopo* tentaculis extensis imaginem florum hortorum referunt; *Eolidia papillosa*, variis coloribus insignis, *Doris tuberculata* rubiginosi coloris, *Polycera quadrilineata* lineis flavis, *Actæon minutum* colore atro distinctum multæque aliæ species, in Laminareis brunneis, in Delessereis, Ptilota, Odonthalia sanguineis, circumnatantes, oculos copiosa varietate coloris formæque pascunt, et *Patella pellucida* splendorem pæne tropicum colore iridis præbet. Paucis igitur his verbis characterem generalem animalium huius regionis describere possumus: *pleraque animalia, hic obvia, splendore colorum plantis, inter quas morantur, non cedentia, varietate etiam pæne superantia, aut tarde tantum vagari possunt*[1]) *aut pæne affixa sunt; flumen enim continuum alimenta adducit, quatenus plantis non vescuntur. Tegumento duro plerumque carent*[2]) *sed violentiæ undarum non exposita sunt.*

Nota. Ex regionibus Francogallicorum et Norvegicorum littorum hæc regio omnino Laminariearum respondet.

---

[1]) Etiam Crustacea huius regionis propria, facultate perexigua movendi prædita sunt (velut Caprella, Pycnogonum cet.).

[2]) Ex classibus et Molluscorum et Echinodermatorum et Polyporum nudæ formæ huius regionis maxime propriæ sunt.

§ 31.

Enumeratio omnium animalium in regione Gymnobranchiorum.

Crustacea.

Caprella linearis Latr. Kullen — Hellebæk — *Leptomera pedata* Latr. Hellebæk. — *Pycnogonum littorale* Mül. Kullen — Snedkersteen. — *Nymphon grossipes* Fabr.? Kullen. — *Phoxichilidium petiolatum* Kr. Kullen — Phoxichilidium femoratum Rathke Kullen — Hellebæk. — Phoxichilidium nov. sp.? Kullen.

Mollusca.

Doris tuberculata Cuv.? Kullen. — *Doris nov, sp.?* affinis D. verrucosæ. Kullen — *Doris obvelata*[1]) Kullen. — *Polycera quadrilineata* Cuv. Kullen — Hveen — *Idalla caudata* nob. Kullen[2]). — *Tritonia velata* nob.[3]) Hellebæk. — *Eolidia papillosa* Cuv. Kullen — Hellebæk. — *Actæon minutum* Sars Hveen. — *Patella pellucida* L. Kullen. — *Patella virginea* Mül. Kullen — Landskrona. — *Patella tessulata* Mül. — *Chiton cine-*

---

1) Zool. Dan. t. 47.

2) Corpore oblongo, multo altiore quam lato, aurantiaco, cauda tenui sursum curvato, 3 laciniis simplicibus in utroque latere dorsi, 6 appendicibus tentacularibus, quarum duabus duplo longioribus, branchiis analibus octo; long 4—5''', lat. 2'''.

3) Corpore oblongo antice truncato postice acuminato, lacteo cum linea laterali flava; duabus appendicibus triangularibus maximis ex margine anterioris corporis partis exeuntibus, reflexis; branchiarum 6 parium secundo duplo maiore quam primo, ceteri sensim minoribus.

reus L. — *Chiton ruber* L. Kullen — Landskrona. — *Ascidia rustica* L. Kullen[1]) Hveen.

Echinodermata.

Cuviera squamata Jæger (Holuthuria squamata Zool. Dan.) Kullen — Landskrona — *Psolus phantopus* Oken (Holuthuria phantopus Zool. Dan.) Kullen — Hellebæk. — *Pentacta pentactes* Jæger (Holuthuria pentactes Zool. Dan.) Kullen — Landskrona. — *Holothuria fusus* Zool. Dan.? Landskrona.

Polypi.

Metridium plumosum Oken, *Actinia plumosa* Zool. Dan. T. 88 inter Hveen et Landskronam. — *Ectacmœa candida* Ehrb., Actinia candida Zool. Dan. T. 115 cum præcedente. — *Isacmœa digitata* Ehrb., Actinia digitata Zool. Dan. T. 133. Kullen. — *Isacmœa coccinea* Ehrb., Actinia coccinea Zool. Dan. T. 63 inter Hveen et Landskronam. — *Isacmœa viduata* Ehrb. Actinia viduata Zool. Dan. Tom. 63 Kullen. — *Sertularia abietina* L. Kullen. — *Plumularia falcata* cum præcedente. — *Plumularia pinnata* Lamk. cum præcedente. — *Tubulipora transversa* Lamk. in Laminaria ad Hveen. — *Crisia eburnea* Lamour. cum præcedente — *Cellaria sp. nov.?*[2]) cum præcedente. — *Flustra foliacea.* L. Kullen. — *Flustra setacea* Kattegattet. — *Flustra pilosa* L. Hveen. — *Alcyonidium gelatinosum* Johnst. Kullen. — *Alcyonidium echinatum* Flem. Kullen. — *Corallina officinalis* L. Kullen — Hellebæk. — *Nullipora polymorpha* Lamk.

---

1) Tres aliæ species huius generis, quas ad Kullen inveni, fortasse novæ sunt.

2) Medium locum inter C. avicularíam et C. scruposam tenet; huic cellularum forma, illi habitu proxima.

— *Spongia coalita, fragilis, ossiformis* et quatuor aliæ speciet, quæ fortasse novæ sunt.

---

§ 32.

Regio Buccinoideorum.

Profunditas.

(*In tabula colore luteo notata.*)

Terminum huius regionis cum ea parte freti, quæ ex indole fundi regio argillacea vocata est, congruit. Totam igitur inferiorem partem convallis, quam fretum inter Scaniam et Siællandiam constituit, occupat. Hæc freti pars regionibus, quæ distributionem solam verticalem indicant, sensu strictiore non adnumeranda est, sed tam parva latitudine est, ut regionis indolem habeat. Et in hac regione animalia aliis externis momentis, atque in ceteris afficiuntur. Fundus enim ex argilla soluta et molli constat, quam facillime penetrare possunt. Itaque animalia in cuniculis vel cavernis variis reperiuntur. Plerumque anterior corporis pars eminet, quia voraces animalibus præternatantibus insidiantur; sunt enim omnia carnivora. Plantæ, ut antea demonstravimus, hic non reperiuntur. Hac igitur re hæc regio a ceteris, quæ tam varia et copiosa vegetatione gaudent, magnopere differt, quæ diversitas ad characterem animalium constituendum magni momenti esse facile apparet. Aqua hac re quoque differt, quod ad fundum maiore tranquillitate est; undæ enim fundum afficere nequeunt, nec unquam impetus fluminis tantus est, quantus in adiacente regione. Splendorem varietatemque colorum, quæ simul cum vegetatione insigni admirationem nostram excitabant, hic non reperimus; varietas autem formarum et fortasse multitudo individuorum hac in regione maximæ sunt. Ex Crustaceis Decapoda huius regionis magis quam ceterarum propria esse videntur; *Stenorynchnus phalangium*, *Hyas araneus*, *Pagurus Bernhardus* frequentissimi sunt. Ex Annulatis *Aphrodita aculeata* in summa tantum profunditate, *ergo* circum Hveen reperitur; *Lepidonote punctata* plerumque in testis vacuis; *Lumbrineris fragilis* una cum *Scoloplo armigero* in fundo maris Lumbricis in terra respondent. Pæne eodem modo *Nephtys borealis et assimilis* et *Glycera alba* obveniunt, *Nereis pelagica* propter magnitudinem, vivacitatem, coloris splendorem a piscatoribus rex vermium (Ormekonge) appellatur. Ex Molluscis inprimis propriæ formæ sunt. *Buccinum undatum* et *Fusus antiquus* per maximam fundi partem tanta multitudine adsunt, ut a piscatoribus ad escam adhibeantur.[1]) *Rostellaria pes pelecani*, *Turritella ungulina*, *Dentalium entalis* argillam profunde penetrant. Ex Acephalis præcipue commemoranda sunt: *Pecten opercularis*, qui quoque in argilla profunde latet; nam etsi testarum eius vacuarum magna multitudo reperitur, rarissime specimen etiam parvulum cum animali vivo invenitur; *Nucula rostrata*, quæ multitudine abundant, maxima celeritate et velocitate pede longo argillam penetrat.[2]) *Modiola vulgaris* præcipue extra Hellebæk,

---

[1]) Ad ea captanda crates, in quo gadus mortuus positus est, in fundum argillaceum demergitur; quem piscem, cui cibus blandissimus ipsi sunt, summa voracitate comedunt. Cum ita crates hisce animalibus repleta est, ad superficiem trahitur.

[2]) Victus horum animalium, quæ in tanta profunditate commorantur, argilla, quæ ex fundo arcessita est, in vasis magnis aqua infusa servanda optime perspici potest. Diu hic vivere possunt, et magna multitudo specierum parvarum, quæ alias oculos nostros effugerent, apparet.

ubi argilla a littore non procul abest, frequentissima est. Itaque a piscatoribus sæpius capta propter suavem saporem laudatur. *Cardium* |*echinatum*, cuius pes ruber longissimus digito simillimus est, in argilla se occultans, radulam facillime evitat; itemque *Cyprina islandica*. Præterea nominanda sunt: *Tellina deprassa*, cuius testa, quamquam est frequentissima, plerumque vacua est, *Hiatella arctica*, quæ bysso aliorum testis adhæret, *Solen pellucidus*, per cuius testam tenuissimam animal pellucet.

Ex Echinodermatis *Spatagus flavescens*, *Solaster endeca* et papposus tantummodo in summa profunditate obveniunt; *Ophiolepis scolopendrica et filiformis* inprimis in testis vacuis vivunt, quibus tantopere adhærent, ut sæpe amovendo dirumpantur. Ex polypis dignissimæ sunt, quæ animadvertantur, *Pennatula phosphorea et Virgularia mirabilis*, cuius maxima pars velut planta ex argilla eminet. Parva specimina motu lento argillam penetrare vidi.

Paucis his verbis generalis character animalium huius regionis describi potest: *plurima animalia, quæ hic obveniunt, carnivora, plus minusve argilla obtecta vivunt. Si testam habent, ea raro magnæ crassitudinis est.*

---

§ 33.

Enumeratio animalium omnium in regione Buccinoideorum.

Crustacea.

Stenorynchus Phalangium Lamk. Hellebæk, Kullen. — *Hyas araneus* Leach. Kullen. — Landskrona. — *Pagurus Bernhardus* Fabr. Kullen —Hveen. — *Galathea strigosa* Fabr. Kullen, Hellebæk. — *Homraus vulgaris*

Edw. Kullen. — *Amphitoe sp. nov.?*[1]) Kullen — *Podocerus Laechii* Kr. Hellebæk. — *Cuma Rathkii* Kr. Kullen Hveen. — *Cuma nasica* Kr. ibidem. *Cuma lucifera* Kr. cum præcedentibus. — *Asellus marinus nob.*[2]) Hellebæk. — *Balanus sulcatus* Brug. præcipue in Modiola.

Annulata.

Aphrodita aculeata Baster Kullen — Hveen. — *Aphrodita Hystrix* Aud. et Edw. Hornbæk Kröyer. — *Lepidonote punctata* Örsd. Kullen — Hveen. — *Lepidonote lævis* Örsd. Hveen. — *Lepidonote assimilis* Örsd. Hveen. — *Lumbrineris fragilis* Örsd. Kullen — Hveen. — *Nereis pelagica* L. — *Castalia punctata* Örsd. Kullen — Hveen. — *Syllis armillaris* Örsd. Kullen — Hveen. — *Eulalia viridis* Sav. ibidem. — *Eulalia sanguinea* Örsd. Hellebæk. — *Eteone Sarsii* Örsd. Hveen. — *Eteone maculata* Örsd. Kullen. — *Eteone pusilla* Örsd. Hveen. — *Phyllodoce Groenlandica* Örsd. Kullen. — *Nephtys borealis* Örsd. — *Nephtys assimilis* Örsd. — *Glycera alba* Örsd. — *Goniada maculata* Örsd. Hellebæk — Hveen. — *Scoloplos armiger* Blainv. Kullen — Hveen. — *Leucodorum coecum* Örsd. Hveen. — *Disoma multisetosum* Örsd. Hveen. — *Cirratulus borealis* Lamk. Kullen — Hellebæk. — *Dodecaceria concharum* Örsd. Hellebæk. — *Ophelia mamillata* Örsd. Kullen. — *Ophelina acuminata* Örsd. Lands krona. — *Eumenia crassa* Örsd. Hveen. — *Chætopterus norvegus* Sars? Hellebæk. — *Serpula triquetra*

---

1) Tentaculis longis circumagendis efficit planum depressius rotundum, ex cuius centro solum caput rubrum, cetero corpore latente, prominet.

2) Unica species huius generis, quæ in mari inventa est. Fortasse proprium genus constituere debet.

L. — *Sabella pavonia* Sav.? Kullen — Hveen. — *Sabella sp. nov.?* — Hellebæk. — *Hermella sp. nov.?* Kullen. — *Terebellides sp. nov.* Hveen. — *Terebella cirrata* Gmel. — *Amphitrite Gunneri* Sars? Hellebæk. — *Chloræma Edwarsii* Dujar.[1]) Hellebæk. — *Pherusa plumosa* Blainv. — *Clymene intermedia* nob.[2]) Hellebæk. — *Clymenia tenuissima nob.* Hellebæk.[3]) — *Lumbricus marinus* nob. inter Hveen et Landskronam. — *Mesopachys marina* nob. Kullen.[4])

Apoda.

Leptoplana atomata Örsd. Kullen — Hellebæk. — *Leptoplana nigripunctata* Örsd. Kullen. — *Typhlolepta coeca* Örsd. Hveen. — *Cephalotrix bioculata* Örsd. Hveen. — *Astemma rufifrons* Örsd. — *Tetrastemma varicolor* Örsd. Kullen. — *Tetrastemma fuscum* Örsd.

---

1) Ann. d. sc. nat. Tom. VI. Pl. 1.

2) Medium locum inter Clymenam amphistoma et Cl. Ebiensem Aud. et Edw. tenet; quoad enim formam capitis huic, quoad anum illi proximum est. Distinguitur: segmentis 24, anterioribus 10 et posterioribus 3 brevissimis, ceteris 4—5plo longioribus qnam latis, capite subgloboso, excisuris analibus triangularibus minutissimis.

3) Charac. gener. Corpus filiforme tenuissimum ex segmentis numerosis distinctissimis constans, longitudine segmentorum latitudinem multum superante, caput clavatum, os terminale, oculi duo minutissimi, cauda depressa. Setæ ut apud Clymenam. Differt a Clymena, cui proximum est, præcipue forma capitis et caudæ, et ore terminali.

4) Genus e Lumbricillorum familia. Corpus fusiforme ex segmentis 24—25 indistinctis brevissimis constans, caput nullum distinctum, os inferum, setarum fasciculi 4 in omnibus segmentis, setis capillaribus. Tubo cibario torto libero, omni constrictione destituto, ab omnibus aliis generibus huius familiæ distinguitur.

Kullen. — *Nemertes assimilis* Örsd. Kullen. — *Nemertes flaccida* Örsd. — *Nemertes pusilla* Örsd. *ibidem.* — *Nemertes maculata* Örsd. Hellebæk. — *Polystemma roseum* Örsd. — *Polystemma pellucidum* Örsd. ibidem. — *Cerebratulus marginatus* Renieri? Hveen. — *Malacobdella grossa* Blainv. (Hirudo grossa Z. D.) in Cyprina Islandica. — *Ichtyobdella sanguinea* nob. Hellebæk. — *Anguillula oculata* nob. Kullen[1]). — *Phascolosoma concharum* nob. (Sipunculus sp. Rathke in Naturhistorie-Selskabets Skrifter V Bind) in Dentalio, Turritella ungulina et. al. Kullen — Hveen[2]).

Mollusca.

Gasteropoda.

Velutina capuloides Blainv. Hellebæk. — *Buccinum undatum* L. — *Fusus antiquus* Lamk. — *Fusus* (*sp.?*) Lamk. Kullen. — *Trophon clathratum* (Murex cl. L.) Kullen — Hveen. — *Defrancia* sp. Kullen. — *Rostellaria pespelecani* Lamk. — *Dentalium entalis* L. Kullen — Hveen.

Acephala.

Pecten opercularis Mül. Kullen — Hveen. — *Pecten striatus* Zool. Dan. T. 60 ibidem. — *Pecten septemradiatus* Mül. ibidem. — *Anomia squamula* L. Kullen — Landskrona. — *Anomia undulata* L.? — *Anomia aculeata* Mül. — *Anomia nov. sp.?* omnes inter Hveen et Landskronam. — *Nucula nov. sp.?* Hveen. — *Nucula margaritacea* Lamk. Kullen — Hveen. — *Leda rostrata* (Nucula rostrata Lamk.). — *Leda intermedia* nob.[3])

---

1) Disinguitur oculis duobus brunneis.

2) A Phascolosomate granulalo Leuchart (Breves animalium descriptiones f. 5), cui simillimum est, distinguitur præcipue corporis parte anteriore tenuiore et multo longiore.

3) Medium locum inter L. rostratam et L. complanatam Mül.

Hveen. — *Modiola vulgaris* (Mytilus Modiolus L.). — *Modiola discrepans* Montg. — *Cardium echinatum* L. Kullen — Hveen. — *Abra tenuis* Leach. Kullen — Hveen. — *Abra nitida* (Mya nitida Mül. ex auctoritate Beckii) — *Psammobia faeroensis* Lamk. Hellebæk — Landskrona. — *Cryptodon flexuosum* Turt. Hveen. — *Tellina depressa* Gmel. Kullen — Hveen. — *Astarte Danmoniensis* Sow. Hveen. — *Astarte striata* Kullen. — *Lucina radula* Lamk. in sinu Codano. — *Cyprina islandica* Lamk. — *Venus gallina* L. *Kullen* — *Hveen.* — *Solen pellucidus* Penn. ibidem. — *Hiatella arctica* Lamk. — *Mya truncata* L. Hellebæk — Hveen.

Echinodermata.

Spatangus flavescens Zool. Dan. — *Spatangus pusillus* Mül. Kullen. — *Solaster papposus* Forbes (Asterias papposa) Hellebæk — Hveen. — *Solaster endeca Forbes* (Asterias endeca) cum præcedente. — *Astropecten arantiacus* (Asterias arantiaca Zool. Dan.) Kullen — Hveen. — *Asteracanthion roseus* M. T. (Asterias rosea) Hveen. — *Ophiolepis scolopendrica* M. T. (Asterias aculeata). — *Ophiolepis filiformis* M. T. (Asterias filiformis) Kullen — Hellebæk. — *Ophiolepis ciliata* M. T. (Asterias ophiura — *Ophiothrix fragilis* M. T. (Asterias fragilis) ubi?

Polypi.

Hydropsis gelatinosa nob.[1]) Kullen. — *Pennatula phosphorea* L. Kullen — Hveen. — *Virgularia mirabilis*

(Krøyers Tidsskrift 4 B. 1 H. p. 90) tenet; illi quoad formam proxima, eodem modo atque hæc, quam magnitudine æquat, sulcata est.

[1]) Hoc distincttissimum genus certe ordinem vel saltem familiam novam, transitum inter Actinias et Hydras conciliantem, con-

Lamk. cum præcedente. — *Pedicellina echinatis* Sars in Dentalio ad Kullen. — *Pedicellina gracilis* Sars cum præcedente. — *Cliona celata* Grant. in Mediola vulg. — *Tethya nov. sp.?* Hveen. — *Acyonium nov. sp.?*

Animalium distributio horizontalis per fretum.

§ 34.

Supra vidimus, salisitudinem aquæ freti sensim a boreali ad australem partem minui, qua re magna diversitas vegetationis partis australis et borealis efficitur. Jam demonstrabimus, animalia eodem modo affici, ut in australi freto magna ex parte cum animalibus aquæ dulcis congruant. In boreali enim parte pæne eædem species atque in sinu Codano reperiuntur; etiam formæ arcticæ hic sunt. Ex solis Maricolis novem species Grönlandiæ et freti communes inveniri iam antea ostendi[1]). Quarum terminum distinctum insula Hveen, quasi agger obiectus flumen boreale excipiens, efficit. Hæc videtur esse causa, cur septentrionem versus et magna ex parte circum Hveen maxima copia specierum in freto obviarum congeratur. Itaque in hac insula brevi commorantes maiorem copiam specierum adipiscemur, quam si aliis locis multo diutius commorabimur. Kullen fortasse solus est locus, qui cum Hveen hac in re comparari possit. Quum circum Hveen

---

stituere debet. Quoad indolem corporis et tubi cibarii Hydræ omnino similis, tentaculis 16 biseriatis, quæ Actiniarum modo contrahi possunt, prædita est. Long. 3‴ lat. 1.‴ Actiniæ proliferæ Sars affinis esse videtur.

1) Conspectus annulatorum Danicorum, auctore A. S. Örsted. p. 8.

Dentalium entalis, Pecten opercularis, septemradiatus et striatus, Nucula margaritacea, Cryptodon flexuosum, Psammobia faeroensis, Astarte Danmoniensis, Cyprina Islandica, Ophelia mamillata, Sabella pavonia, Pennatula phosphorea, Virgularia mirabilis et m. al. obveniant, inopia, quæ mari Baltico propria est, iam spatio dimidii milliarii septentrionem versus a Hveen apparere incipit, ut australis freti pars cum mari Baltico omnino congruat. Paludinellæ hic dominantur; huc species generum, quæ aquæ dulcis propriæ sunt, accedunt, velut Limnæa Baltica, Neritina Baltica; etiam species, quæ alias tantummodo in aqua dulci inveniuntur, velut Planaria lactea et torva. Præterea magna copia larvarum Insectorum, præcipue Dipterorum, hic obvenit.

Nota. Multæ species, quæ per maximam partem fundi fere eodem modo distributæ sunt, nonnullis locis tanta multitudine coacervatæ apparent, ut hæ propriæ earum sedes habendæ sint. Quamquam habitationes omnium specierum et rararum et vulgarium accurate cognoscere magnopere interest, hic tamen habitationes tantum specierum, quæ singulis locis frequentissimæ obveniunt, indicabo. Quum ex nonnullis specimina pauca vel etiam sæpe mutilata repererim, non dubito, quin huius rei causa in mancis investigationibus quærenda sit, et quin loca inveniantur, quibus pleræque frequentiores sint. Hæ species magna multitudine iis locis, quæ infra in tabula prima litteris notavi, reperiuntur: *Pagurus Bernhardus* ad Hellebæk (a). *Lumbricus fragilis* ad Hveen (b). *Nereis pelagica* ad Hellebæk (c). *Nereis diversicolor* extra Clasens Have (d). *Nephtys borealis inter* Hellebæk et Viken (e). *Scoloplos armiger* inter Viken et Höganæs (f). *Spio seticornis, Jæra albifrons*, *Corophium longicorne*, *Monocelis lineata* inter se commixtæ obveniunt in Kallebodstrand ad Ny-Badehuus (g). *Planaria Ulvæ et Nemertes olivacea* inter Havniam et Trekroner (h). *Patella pellucida* ad Kullaberg (i). *Turritella ungulina* et *Rostellaria pes-pelecani* inter Gilleleie et Kullen (k). *Dentalium entalis* inter Sletten et littora scanica (l). *Pecten opercularis* inter Hellebæk et Viken (m). *Nucula rostrata* ad Hveen (n). *Modiola vulgaris* ad Hellebæk (o). *Modiola discrepans* ad Hveen (p). *Mactra nov. sp.?* ad Mölle (Kullen) (q). *Astarte Danmoniensis* ad Hveen (r). *Cyprina Islandica* inter Hveen et Scaniam (s) *Mya arenaria* et *Cardium edule* in vado Disken (t) et in Kallebodstrand (t). *Psolus phantopus* ad Hellebæk et ad Kullen (u). *Echinus esculentus* inter Skovshoved et Scaniam (v). *Spatagus esculentus* inter Sletten et Hveen (x). *Ophiolepis filiformis* ad Kullen (y). *Virgularia mirabilis* et *Pennatula phosphorea* ad Kullen (ö).

Nota. 2. Numerus specierum animalium omnium (Infusoriis exceptis) adhuc in freto inventarum est 427. Ita inter classes diversas distributæ sunt :

Pisces	90
Crustacea	77
Arachnidæ	3
Vermes	110
Radiata	58
	338.

De momentis externis regiones animalium constituentibus.

§ 35.

Quomodo indolem fundi aqua constituat, præterea quomodo vegetatio et aqua et fundo afficiatur, supra vidimus; iam facile intelligimus, distributionem animalium per regiones diversas non solum aquæ et fundi, sed etiam vegetationis indole constitui. Quomodo animalia ex plantis et plantæ ex momentis externis anorganicis pendeant, nunc demum plane perspicere possumus, quum viderimus, ubi alterum horum mutetur, alterum quoque mutari. Regio enim arenaria et algarum olivacearum et viridium et Trochoideorum regio, item regio lapidaria et regio Laminariearum et Rhodospermearum et Gymnobranchiorum, denique regio argillacea et regio plantis nudata et Buccinoideorum inter se congruunt. Quomodo omnes inter se coniunctæ sint, si ita perspexeris, tum demum vera vis singularum apparet. Si igitur hoc solum cognitum est, animal quoddam in profunditate orgyiarum undecim in freto inveniri, condiciones, sub quibus vivit, facile apparent. In hac enim profunditate fundum esse argillaceum et plantis omnino destitutum scimus. Sic demum quomodo momentis externis organismi afficiantur, quod physiologiæ magni momenti est, intelligere possumus. Cuius rei quæstionem diligentiorem ad aliud tempus differens, tantum indicabo, plurima animalia inferiora marina externis momentis non minus quam plantas affici, ut iisdem terminis definiri posse videantur, quibus plantas Unger definiverit (bodenstete, bodenholde, bodenwage). Hic vero hoc obstat, quod una cum diversitate fundi cetera momenta externa mutantur, ut sæpe difficile sit constituere, quibus animalia maxime afficiantur. Itaque si quoddam animal non nisi in fundo argillaceo reperitur, dubium est, utrum indoles argillæ an indoles lucis, quæ profunditate maiore mutatur, an pressus aquæ, an tranquillitas magna maioris momenti sit, et sæpissime posterioribus momentis animalia plus quam ipso fundo affici intelligentur. Ex hoc schemate interior regionum conjunctio intelligitur.

Profunditas	0—50 ped.	— 50—65 ped.	— 65—? ped.
Fundus	arenosus	— lapidarius	— argillaceus
Plantæ	Chlorospermeæ, Melanospermeæ, Zostera marina.	—Rhodospermeæ	—nullæ
Mollusca	Trochoidea	—Gymnobranchia	—Buccinoidea

§ 36.

Besultata generalia investigationibus freti acquisita.

1) Fretum aut hac periodo ineunte, aut tertiaria exeunte ortum est.

2) Si hac terræ periodo diluvie formatum est, Sjællandiam et Scaniam una tantum vel duabus lingulis conjunctas fuisse veresimile est.

3) Si oræ, quæ fretum cingunt, post formationem plutonicam vim subierunt, parvula elevatio littoris Sjællandici et Scanici fuisse videtur.

4) Formam freti, postquam ortum est, graves mutationes subiisse, non verisimile est. Ad littus Sjællandicum præcipue *alluvionibus*, ad Scanicum præcipue *abluvionibus*, effectæ sunt.

5) Indoles oræ Scanicæ inter Glumslöf et Landskronam declarat, formationem peculiarem argillæ glaucæ plasticæ, fissilis, fossilibus omnino destitutæ, quæ a Forchhammero et in Jutlandia (prope Fredericiam) et in Fionia et in Sjællandia detecta est (Kröyers Tidsskrift 1 B. p. 209), etiam in Scania inveniri.

6) Tria sola genera fundi in freto sunt, arena, argilla, testæ calcareæ (aut lapides congesti); omnibus enim locis, quorum profunditas maior 8—10 orgyiis est, fundus ex argilla, ubi minor est, ex arena constat. Testæ calcareæ aut lapides congesti non nisi, ubi et arena et argilla flumine ablutæ sunt, reperiuntur.

7) Hæ rationes nisi respiciuntur, vera unitas varietatis formationum geologicarum, quæ normales vocantur, intelligi non potest.

8) Unaquæque enim formatio geologica ex tribus constat membris, quorum primum arenæ, secundum argillæ, tertium testis calcareis respondet.

9) Ergo natura in terra formanda eandem legem

secuta est, quæ in regnis organicis valet, evolutionem per series parallelas cum membris analogis.

10) In formatione tertiaria arena (Rullesteenssandet) vadis, argilla (Rullesteensleret) locis freti profundioribus respondet.

11) Ergo hæc duo membra, in quæ formatio tertiaria Daniæ a Forchhammero divisa est, *eodem tempore* orta sunt.

12) Si Dania diluvie ab oriente profecta inundata esset, fundus locornm inundatorum ex arena, non ex argilla constaret.

13) Ex fundi indole, si ex massis solutis constat, de stratis terræ, quæ in eadem profunditate sunt, nihil certi coniici potest.

14) Algæ freti in tres regiones distributæ sunt, quarum cuique singuli subordines, in quos hic ordo dividitur, Chlorospermeæ, Melanospermeæ, Rhodospermeæ, proprii sunt, ut Rhodospermeæ in maxima Melanospermeæ in minore, Chlorospermeæ in minima profunditate obveniant.

15) Hæc distributio præcipue constituitur ea ratione, qua radii lucis varii, mare penetrantes franguntur, quum radii rubri maximam profunditatem attingant, hic vero algæ purpureæ obveniunt, deinde aurantiaci cett., ut cærulei minime penetrent; Oscillatoriæ vero viridi-coerulescentes in superficie maris dominantur.

16) Peculiare illud, quod toti subordines algarum eodem colore sunt, vel quod idem est, quod color, qui alias character exigui momenti nota tantum specifica est, hic ordines constituit, eo explicatur, quod algæ solæ plantarum luce, ex qua color semper pendet, tam varie afficiuntur.

17) Et in algarum ordine evolutio per series parallelas cum membris analogis fit.

18) Pleraque animalia marina inferiora externis mo-

mentis ita afficiuntur, ut velut plantæ certis stationibus adstricta sint.

19) Ex quo fit, ut animalia marina certis regionibus referri possint.

20) Mollusca optimum fundamentum divisionis præbent.

21) Secundum mollusca tres regiones, *Trochoideorum*, *Gymnobranchiorum*, *Buccinoideorum* constitui possunt.

22) Hæ regiones partim diversa vegetatione partim diversitate momentorum externorum, quæ diversa profunditate constituitur, nituntur.

Explicatio figurar. xylograph. et tabul.

Fig. xylogr. 1 p. 8 Radula Ballii.

— 2. Collis ad Villingebæk transsectus. a strata arenæ vagæ. b stratum animalium marinorum.

— 3. Strata, quæ, fossa ad Frederiksleie prope Landskronam facta, apparent. a humus arenosus, b stratum animalium marinorum, c stratum argillæ cum ferro oxydato commixtæ, d stratum lapidum, e argilla lutea vulgaris.

— 4. Prærupta inter Sletten et Humlebæk. a argilla lutea vulgaris, b strata argillæ mollissimæ absque lapidibus, c strata lapidum, d conglomeratum Calcarei.

— 5. Prærupta ad Glumslöf in Scania; a argilla lutea vulgaris lapidibus commixta; b strata humi, c strata calcis aquæ dulcis; d argilla glauca plastica fissilis.

Tab. I—II.

Fig. 1—6. Sectiones transversales freti, quæ diversam profunditatem in diversis freti partibus declarant. Numeri additi indicant profunditatem secundum profunditatem. In tabula prima colorl uteus significat argillam et regionem Buccinoideorum, color ruber regionem Rhodospermearum, color intense brunneus regionem Laminariearum, uterque regionem lapidariam et regionem Gymnobranchiorum; color dilute brunneus regionem Melanospermearum, color viridis regionem Chlorospermearum, coeruleus regionem Oscillatorinearum; hi tres colore significant quoque regionem Trochoideorum et regionem arenariam. Tabula secunda ostendit, quomodo hæ regiones continuatio regionum montanarum haberi possint. Hic colores diversi idem atque in tabula prima significant.

Corrigenda.

Scriptum est:	Scribi debet:
p. 13 6 v. im. Iul.	Iun.
- 32 12 v. membronum	membrorum.
- — 13 v. formatiorum	formationum.
- 41 8 v. Conferva distans Dillv.	Conferva diffusa Dillv.?

Addenda.

§ 12 Nota 1 indicatur argillam tertiariam fossilibus omnino destitutam esse; me enim fefellit ea strata formationis tertiariæ, quæ a Forchhammero in insulis: Langeland, Æröe, Als et al. detecta erant (v. Oversigt over Vidensk. Selsk. Skr. 1842 Nr. 5) ad hanc mediam formationis tertiariæ partem pertinere. Qua observatione sententia mea miro modo confirmatur, nom Cyprina Islandica in ea argillæ tertiariæ parte dominari indicatur, et antea cognitum est in arena tertiaria præcipue Cardium edule, Mytilum edulem, Nassam reticulatam dominari; in mari vero hujus periodi Cyprina Islandica profunditatis ɔ: argillæ, Cardium edule et al. vero locis maris minus profundis ɔ: arenæ propria sunt, ut analogia maxima inter hujus et illius maris fundos sit.

Tab. I.

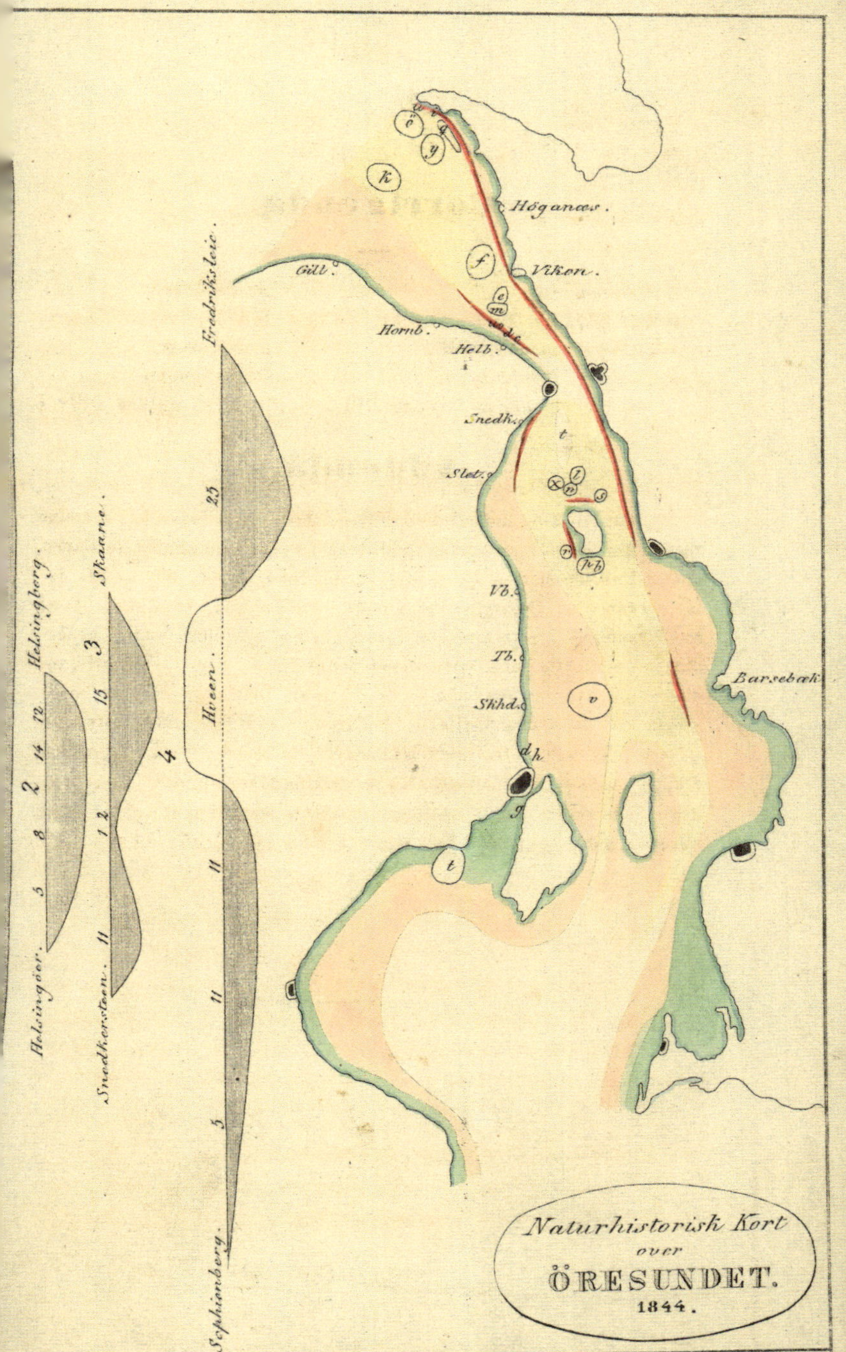

Naturhistorisk Kort
over
ÖRESUNDET.
1844.
Höganæs.
Vikon.
Gill.
Hornb.
Helb.
Snedk.
Slet.
Vb.
Tb.
Skhd.
Barsebæk
Helsingöer.
Helsingborg.
Snedkersteen.
Skaane.
Hveen.
Sophienberg.

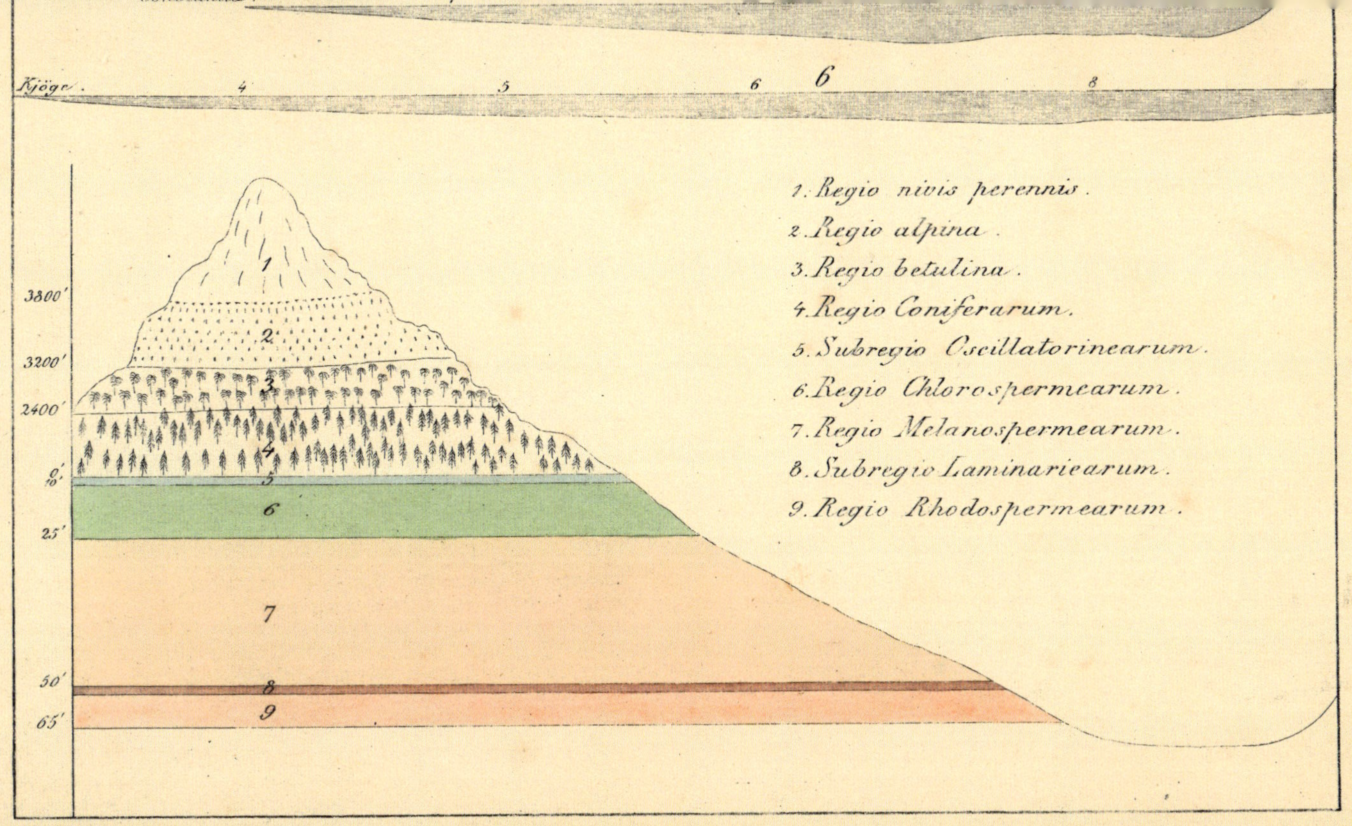
Kjöge.
4
5
6
6
8
3800′
3200′
2400′
25′
50′
65′
1
2
3
4
5
6
7
8
9
1. Regio nivis perennis.
2. Regio alpina.
3. Regio betulina.
4. Regio Coniferarum.
5. Subregio Oscillatorinearum.
6. Regio Chlorospermearum.
7. Regio Melanospermearum.
8. Subregio Laminariearum.
9. Regio Rhodospermearum.

ABOUT THE AREAS OF THE SEA.

Elements for the Sound Natural History Topography.

Inaugural dissertation,
as A.S. Ørsted, Cand. Phil. has written
with a view to acquiring dignity
as a Master of Arts and will try to defend
in public on April 29 with the venerable
E. Petit, military surgeon, as respondent.

COPENHAGEN
Printed by J. C. Scharling
1844

The Faculty of Philosophy at the University of Copenhagen considers this thesis worthy of being subjected to the public examination of the scholars in order to obtain the master's degree in accordance with applicable rules.

I. Reinhardt,
this year's Dean of the Faculty of Philosophy

To my eminent, much-loved uncles

A.S. ØRSTED,

who from my earliest childhood
has been in my very thoughtful
father's place, and

H.C. ØRSTED,

who has always been the eminent
supervisor in my studies, both of whom
I have always looked up to as excellent
examples of everything good, exalted
and excellent, the author has wished
this thesis to be dedicated.

The author

Table of contents for the little book.

a) The coasts of Zealand.

b) The coasts of Scania.

c) The special formation of plastic blue clay that can be split [varved clay], in Scania. § 8 note.

Description of three geological areas. § 9.

Explanation of the three geological areas. § 10.

The results from the previous section transferred to geology. § 11–13.

a) On the geology of Denmark. § 12.

b) On the universal geology. § 13.

CHAPTER 3.

The botanical conditions or the areas of algae of the Sound. § 14–25.

What has previously been written about the areas of algae. § 14.

About the three areas of the Sound corresponding to the subordinate groups of the algae. § 15.

About the area of green algae. § 16.

Enumeration of all the previously known green algae in the Sound. § 17.

About the area of brown algae. § 18.

Enumeration of all the previously known brown algae in the Sound. § 19.

About the area of red algae. § 20.

CHAPTER 4.

The zoological conditions or the areas of the animals of the Sound. § 26–36.

PREFACE.

As I often wandered along the shores of the Øresund Strait, delighting in the varied character of the land and its luxuriant vegetation, it seemed to me a most worthy endeavour to dispel, if possible, the obscurity that shrouded so vast a portion of the Earth's surface, where the sea's unbroken expanse concealed from view the manifold wonders of the ocean floor and the countless exquisite forms of life hidden beneath its depths. Although the Sound for me already presented animals and many new and rare plants, however, I was only partly satisfied with this; there was a lack of a holistic picture, which will only be present when the individual parts are so connected that we understand the whole, and the unity also shows itself in the diversity. I assumed that numerous organisms are not randomly spread throughout the depth of the ocean but are distributed according to certain laws. Thus, since it had become possible for me to allocate the whole of the year 1842 to these studies, which my mind had long burned to deal with, the fruits which I collected by a methodical examination of the sea did not disappoint me. What I myself, by examining the Sound, have understood regarding the conditions in general, I have dared to inform others with this thesis, which contains elements of the natural topography of the

Sound. This little book is entitled 'About the regions [zones] of the sea', because the whole geographical system of animals and plants appears as areas, although the Sound is so narrow that it consists of only two sloping sides. The term 'area' is here used, as it is used by most authors today. The extension is vertical [depth distribution]. In the case of the geographical condition of animals, this term denotes the same. The fact that the previous geography of the plants has been brought into a system is so useful and fruitful to the geography of the animals that we can find from it not only the outline of the whole system, but also many individual areas, which today are accepted by all. The geography of the plants has therefore in a wonderful way just prepared the geography of the animals.

I had first decided to divide this little book into two parts, of which the first should contain general considerations, the second as complete as possible an enumeration of the plants and animals found in the Sound, specifying habitats and description and depiction of new species, which are not few. But for fear that the descriptive part would be larger than would be suitable for a dissertation, I decided it was better to add an enumeration that was not more comprehensive than necessary in order to allow to compare the organisms in the Sound with others, and simply include brief descriptions of the new species that need to be described. Perhaps we can supplement what is missing elsewhere.

It is a great pleasure for me to thank the famous gentlemen Hofmann Bang, I. Agardh, Liebmann and the doctors Beck and Kröyer, because they have in many ways most carefully helped me with identifications of many species, the first three of algae, the last two of animals.

Only those who are engaged in exploring nature can understand the very great pleasure that I got from exploring the Sound, as the individual parts began to grow together and become unified, and the unity of the different parts emerged more clearly. Therefore, I will always remember the time when I visited the Sound as something very delightful in my life, and I wish with all my heart that it will be possible for me to explore the other seas that surround the Danish islands as well.

On the lists, the following should be noted:

1. If no habitat is specified for a species, it means that it is common in most of the area to which it is associated.
2. If two places are mentioned with a short line in between (such as Kullen-Landskrona), it indicates the species' extreme northern and southern occurrence, respectively.
3. Where the name of the finder of a species is omitted, I have found it myself at the place indicated.
4. A few species that I have included from places outside the Sound doubtless occur in the Sound as well.

HISTORICAL INTRODUCTION.[1]

Not all forms of science can evolve at the same rate. A superficial knowledge of the history of the sciences shows that a number of them have developed slowly, and in this way, one has always prepared another and paved the way for another, and it could not be otherwise. A proof of this is science that teaches us the laws that govern the connection plants and animals have with the surface of the earth, or the doctrine of how their habitat and distribution correspond to their external conditions by which they are affected, i.e. the science that deals with the geography of plants and animals.[2] Until the climatology and physiology of animals and plants reached a certain degree of perfection, and before studies of fauna and flora provided a greater knowledge about the distribution of organisms throughout the world, these sciences could not at all exist. Thus, in the latter part of the last century, a premonition of the geography of plants began to show itself through travels and floras (first and foremost Tournefort's *Voyage au Levant* and Linnaeus' *Flora Lapponica*). In 1807, the real

---

1. Although only a small part of the geography of the animals is highlighted in this little book, I have not thought that it could be otherwise than that I should anticipate a historical overview of the development and current status of this science.

2. Schouw, *Grundtræk til en almindelig Plantegeographie* [Fundaments of a General Phytogeography], p. 5.

science finally emerged in Humboldt's "Essai sur la géographie des plantes" and in his "Tableau physique des régions équinoxiales". Many excellent aids to this science followed, in particular Wahlenberg's *Flora Lapponica*, Robert Brown's "General remarks on the botany of terra australis" in 1814, Humboldt's thesis "Sur les lignes isothermes" in 1817 and Decandol's in *Memoires de la societé d'Arcueil*, just as our compatriot, the learned Professor Schouw in 1822, was able to truly establish this system of science.[1] While in this way the geography of plants has long been a special science, this cannot be said today about the geography of animals; although many and excellent means have been developed, they have not yet been put into a system. So, there seems to be a great lack of works on fauna. As far as the higher animals are concerned, however, this deficiency has been remedied in recent times, but by no means when it comes to the lower ones.

Zimmermann[2], as the first, has brought weighty aids to the geographical distribution of animals and (in the third volume) provided a general survey of the subject in such a manner that nothing more correct could at this time be justly expected. With Treviranus[3] there is a large amount of material concerning many things, but the studies that preceded it were too few to form the basis of a doctrine that

---

1. Schouw op. cit.
2. *Geographische Geschichte des Menschen und der allgemein verbreiteten vierfüssigen Thieren mit einer zoologischen Weltkarte*, vols. 1–3. Leipzig (1778–83).
3. *Biologie*, 2 volumes. Göttingen (1803).

will also be able to satisfy future generations. Illiger in 1811 wrote (in *Act. Academ. Berolin*) excellently and plentifully about the distribution of mammals and birds. A chapter in Rudolph's *Anthropologia* (1812), entitled "Ueber die Verbreitigung der Thiere" [On the distribution of animals], simply shows that it cannot be argued that the animals were originally gathered in one place, whence they were dispersed to widely different parts of the globe. Latreilli's remarks on the geographical distribution of insects[1] is a major step forward in the development of the laws on the general distribution of animals. Fabricius has already tried to divide the earth into eight zoological realms according to the different fauna of the insects. Lesson has tried to supplement Illiger's small work about birds, and Minding's writings on mammalian distribution[2] are only a supplement. Swainson has also contributed a great deal to our knowledge about the division of the land into zoological realms.[3] Based on birds, he divides the earth into six realms: a European, an Asian, an American, an African, an Australian and an Arctic. The preface of Prichard's book on man[4] contains the best description so far of the geographical distribution of the animal kingdom. The geographical distribution of the individual classes of animals is treated, and the

1. *Introduction à la géographie général des Insectes et des Arachnides.* Mémoire du Museum d'histoire naturelle (1815).
2. *Geographische Verbreitung der Säugthiere* (1829).
3. *A Treatise on the Geography and Classification of Animals.* London (1835).
4. *Researches into the physical history of mankind* (1836).

earth is first and foremost divided according to mammals and reptiles which must in fact provide the best foundation for such divisions in these nine zoological provinces: the Arctic realm, the temperate American, the temperate European and Asian, the tropical and the southern African, the tropical Asian, the Indian archipelago, Polynesia and Australia. These general laws are presented for the distribution of terrestrial animals: 1) the Arctic zone, in which three parts of the globe comprise one area, is the only one that has the same species; 2) in the other realms separated by seas, the same species are never found, but sometimes the same genera, and frequently analogous genera and families; 3) there are different species in the two hemispheres; 4) small islands located far from a continent, contain no native mammals; 5) islands located close to a continent contain the same mammalian genera as the continent. From this, it is concluded that animals naturally occur in areas that fit their way of life and have not migrated from the same place, as Linnaeus assumed. Schlegel[1] has given a description of the geographic distribution of the snakes and a map which is a great step forward, not only because it almost exhausts the subject, but also because the distribution over the earth of these animals is more suit-

1. *Essai sur la physionomie des serpens*, vol. 1. Amsterdam (1837), p. 195.

able than the others in order for elucidating the question of the original habitat of the animals, as it is more difficult for them to leave the land they inhabit, there is no obvious reason why humans should move them away from there, and they are not easily exterminated. In the second volume of Lacordair's *Introduction à L'Entomologie Paris* (1835), the geographical conditions of the insects are extensively described. Lyell, in *Principles of Geology*, vol. 3, 5th edition (1837), states many remarkable things to illuminate the distribution of animals. Oswald Heer has listed several things of great importance to illustrate the extent to which the color of animals is affected by the climate. In the Berghaus' "Allgemeine Länder- und Völkerkunde 3 Band 1838" there is a comprehensive but unstructured collection [of information] on the geography of the animals. R. Wagner added in a translation of Prichard's "Physical history of mankind" a summary of Swainson's and Schlegel's books, which I have mentioned above.

But when it comes to determining the geographical distribution of marine animals, there is in fact a lot missing. Zimmermann has correctly said that, because the temperature varies to a lesser degree, there is not the same diversity of animals between the different latitudes as there is on land, but rightly a little too boldly he claims "ein Seethier ist oder könne doch in Meer fast all Orten zu Hause seyn" [a marine animal is, or could occur, in almost all areas] (vol. 3, p. 218). This misconception is primarily due to the fact that analogous species at both poles were assumed to be identical. Peron and Lesueur were the first to point out that this was wrong, showing that twenty different species have been described under the name *Phoca ursina*, and that the same lack of order can be found regarding *Phoca vitulina*. They also demonstrate that Steller and Fabricius have described very different species under the name *Phoca leonina*, and they have taught us better about the distribution of marine animals, asserting that there are indeed analogous species in the Arctic and Antarctic seas, but that they are not identical ("de tous les animaux que nous avons pu voir nous-même, il n'en est pas un seul qui ne se distingue essentiellement des espèces boréales analogues" (*Annales du Museum d'hist. Nat.* vol. 15, p. 300 (1810))). However, they have also tried to maintain the old and very mistaken view of the habitat of corals, reinforcing and further developing the view that many islands in the South Sea were formed by corals (*Voyage II*, pp. 165–92). Quoy and Gaimard have earned great credit because they first rejected that opinion and have shown that corals, with a hard shell of low thickness covering the rocks, cannot be found below the depth of a few fathoms, and that the same animals are unlikely to be able to live in the variety of ocean pressure in which they would have had to live if they should have formed islands from a depth of 1000–1200 feet below the surface (*Annales des sciences naturelles* (1825)). Since then, Ehrenberg in his investigations in the Red Sea has confirmed the truth of these observations, where he everywhere discovered that a thin, hard coral covering of not more than 1–4 feet is formed, and that the same is rarely found deeper than 18 feet and therefore it is unable to form islands at all, but can often protect them from the very dangerous effects of waves.[1] Milne Edwards, who was the first to try with greater accuracy to determine the diversity of animals at different depths of sea, distinguished four areas.[2] Sars has reached the same opinion during his studies of the coasts of Norway.[3] Ehrenberg has extensively elucidated the geographical distribution of ciliated animalcules.[4] Milne Edward's book on crustaceans covers extensively everything written about any class of marine animals.[5] By dividing the sea into three realms according to the crustaceans, he concludes, based on his investigations, these general results: 1) the number of species increases from the poles to the equator; 2) the most perfect forms are found in the tropical seas; 3) the amount of individuals is greatest in the Arctic seas; 4) the origin of species seems to come from different centers and not from one place, as Linnaeus assumed; 5) large oceans seem to separate and restrict the distribution of animals. Thus, no species are common to

---

1. 'Ueber die Natur und Bildung der Korallenbänke des rothen Meeres.' *Act. Acad. Berol.* (1832).
2. *Annales des sciences natur.* vol. 21, p. 326.
3. *Beskrivelser og Iagttagelser over nogle mærkelige Dyr* o.s.v. Bergen (1835), p. vi.
4. *Geographische Verbreitung der Infusionsthiere.* Monatsbericht der Berliner Academie (1840), pp. 157–97.
5. *Annal. des sciences natur.* vol. 10 and in *Histoire naturelle des Crustacés*, vol. 3, p. 555.

the tropical coasts of each side of the Atlantic. 6) Long coasts, large series of islands, and water currents are favorable for the dispersal of species. In recent times there has been disagreement among us about the geographical distribution of the crustaceans, since Eschricht, in order to defend the old view, has accepted the opinion that several crustaceans are widespread throughout the entire earth,[1] a view Krøyer has endeavored to disprove.[2]

---

1. *Förhandlingar vid de Skandinaviske Naturforskarnes tredje Möte i Stockholm* (1842), p. 203.
2. *Krøyer Naturhistoriske Tidsskrift* ser. 2, vol. 1, part 5, p. 474.

CHAPTER 1.

The Sound's general physical-geographical conditions.

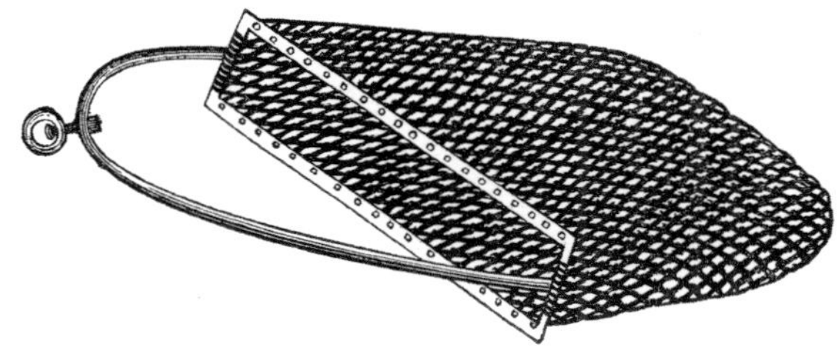

Fig. 1.

About the method of investigation.

§ 1. When I wanted to learn very closely about the Sound and the geographical distribution [zonation] of the animals and plants that occur in it, and there were no other parts of the bottom than those adjacent to the shore that could be useful for my observations, it was necessary, in order for me to be satisfied with my opportunities, to use special instruments and a special method to study the larger part of the Sound. How helpful it would have been to explore these things, to carry the same nature as the excellent swimmer *Piscicola* and walk around on the sea floor just like on land or at least to dive in a diving bell! Since this is not possible, the Ball's dredge is most suitable for sampling things from the bottom that cannot be seen with the eyes. A sketch of this is here added, in which it can be seen that it is constructed in almost the same way as an instrument for collecting oysters.[1] It consists of a rectangular iron frame equipped at the top with an iron bow attached to the two shorter sides, and at the bottom with a net whose heavy threads are joined together into dense meshes. The center of the bow has a small ring that can move around a small shaft. A rope is attached to the small ring. The submerged dredge hangs from a small boat moving slowly forward, just scraping the bottom. In this way, when a part of the bottom has been dredged it gradually becomes filled with animals and plants that occur in it. If a bottom of greater depth is to be examined, lead masses that increase the scraper's weight are attached to the outermost sides of the arch. It must always be towed against the current so that nothing will prevent it from being filled. A triangular or ring-shaped dredge of the same construction is most suitable if one wants to penetrate to the deeper parts of the bottom. But it is not always easy to use this very simple equipment because the current on the bottom is often one and that on the sea surface another, and because some practice is needed so that one can feel if the equipment stays at the bottom to a sufficient degree or too little, which very often can depend on the length of the rope. Therefore, one would not regret having consulted the fishermen who are used to pulling nets back and forth in the depths.

Realizing that it would be impossible, when the dredge was randomly lowered without plan and method, to find

1. H. Krøyer, *De danske Østersbanker*, p. 74 table I.

the true laws that embody the nature of the distribution of organisms in the depths, I propose a safe plan and method to explore the Sound.

Having determined lines that cross the coast of Zealand to Scania from the northern border of the Sound to the southern, I examined the nature of the Sound in as many places as possible. In each place I have recorded whether the bottom was sandy or clayey, and what animals and plants were found there. The location of any place where something special appeared was very carefully marked with a compass and with the fishermen's markings on the coast, which can be very easily observed anywhere on the Sound; for this, the fishermen, who are very experienced, provided excellent help. A geographical plate that I will later try to outline is based on this study of many places. (Plate I)

About the boundaries, division and depth of the Sound.

§ 2. The Sound is the largest of the straits by which the Skagerrak-Kattegat is connected to the Baltic Sea. It lies between 55° 20' north and 56° 18' north and runs from south to north. The northern border is a line from Gilleleje to Kullen, the western Zealand coast is from Gilleleje to Stevns, the eastern Scanean coast is from Kullen to Falsterbo, and to the south it is limited by a line from Stevns to Falsterbo. The maximum length is approximately thirteen miles,[1] the largest width between Copenhagen and Malmö being approximately four miles, and the smallest between Kronborg and Helsingborg being 6340 ells.[2] It is divided into three parts: one southern, one in the middle and one northern. *The northern part* extends from the lowermost part of the Skagerrak-Kattegat to Helsingør and Helsingborg, which is almost the same as the bay of Skagerrak-Kattegat, some of the Sound can therefore also be called the Skagerrak-Kattegat part of the Sound. *The middle part* is delimited to the north by a line drawn between Helsingør and Helsingborg, and to the south between Copenhagen and Barsebäck. Enclosed from so many sides, the nature of the surrounding coasts is almost identical to the bay. *The southern part* extends from the boundary of the middle part to the extreme of the Sound and corresponds entirely to the Baltic Sea and can therefore be called the Baltic part. The middle part is thus the Sound itself, while the other two seas, which are connected to the Sound, are a kind of precursor to it.

The overall depth of the Sound is so shallow that it cannot be described in proper proportion to the width when represented by sections from side to side unless they are set with a very large scale. In the sections shown in Tables I and II, in which two fingers[3] correspond to one mile, the depth-to-width ratio cannot be properly exposed; thus,

1. A Danish mile is 7,532.48 m.
2. 1 cubitus = 1 ell = 0.6277 m. So, 6340 ells = 3,979.618 m.
3. Digitus = finger = 18.5 mm.

seven fathoms should merely be a 571th fraction of two fingers, which is an invisible length. Here seven fathoms are equal to a line, so that only the relative ratio of the depth of the different places of the Sound are shown, but the depth of all these places is very great compared to the width. These sections show that only the middle of the eastern part of the Sound has a greater depth, but that it is the same both in the northern and in the middle parts (Plate I, 2–4; Plate II, 5), in particular towards Sweden; the depth also inversely corresponds to the width, so that in the widest part of the Sound it is not deeper than eight fathoms in the middle[1] (Table II, 6); in the northern part the greatest depth is fifteen fathoms (Plate I, 1); in the middle and narrowest part between Hven and Scania the greatest depth is twenty-five fathoms (Plate I, 4).[2]

About the chemical composition and density of the water.

§ 3. The different parts of the Sound show very large differences in the density of the water which corresponds to the amount of salt dissolved in it. This difference between the southern and northern parts seems to be the same as between the Baltic and the northern seas [Skagerak], and therefore I think that the examination of

1. Orgyia = fathom = 1.84 m.
2. cf. *Speciel-Kort over Sundet* (1840), sea chart originally by Admiral Klint.

the Sound itself can easily be obtained by examining the specific weight of both seas. Marcet presents these results:

Baltic Sea

*Density**	*Depth*
1.0061	surface
1.0171	seventeen fathoms
1.0272	fourteen fathoms

[* where 1 is the density of distilled water]

The same author mentions these relations between the southern Atlantic and the Baltic. The water of the former has a density of 1.02888 and contains 4.26% salt, but the water of the latter has a density of 1.00490 and just 0.66% salt.[1] Thus, the difference in the specific weight of these seas is 0.02396, and regarding the salt it is 3.60; we can confidently state that there is the same difference between the southern and northern parts of the Sound. The difference in the salt content of the water at the same location in the Sound thus also depends on the direction of the flow, so that it can be observed very easily by tasting it. It would not be fruitless and unprofitable if someone would carefully examine the difference in the specific weight of the water, in both the northerly and southerly currents (for example at Copenhagen).

1. *Gehlers physicalisches Wörterbuch*, volume 6: *Artkl. Meer.*

About the temperature of the water in the Sound.

§ 4. Observations of the water temperature in the Sound in the past, measured at the fortress Trekroner, are available in the Act. Acad. Reg. Harbor. Scient. Therein, only the mean temperature is indicated; therefore, from these observations we have calculated the mean temperature for both the entire year 1842 and the individual months, so that we can compare it with the mean temperature of the air this year.

[Month]	The air temperature [°C]	The water temperature [°C]
Jan.	1.4	1.8
Feb.	0.2	1.2
Mar.	2.5	2.3
Apr.	5.4	4.2
May	10.2	9.5
Jun.	12.2	11.5
Jul.	12.9	13.4
Aug.	15.7	16.5
Sep.	11.2	13.0
Oct.	6.3	8.0
Nov.	1.4	3.8
Dec.	3.0	3.8
All year	6.8	7.4

From this one can see that the whole year's water temperature is 0.6 °C higher than the air temperature, and therefore the water temperature is higher [than that of the air] in winter; in the months of March to June, the air temperature exceeds that of the water.

About the current.

§ 5. The currents observed in the Sound are localized, i.e. they are not directly related to the entire oceanic circulation. The nature of the current is the same as the narrow straits, connecting a confined sea in which a quantity of fresh water flows together, often have with the larger seas surrounding it. If, therefore, no other circumstances at the same time affect the direction of the current, a slight current always carries water from the Baltic into the Skagerrak-Kattegat, so that the surface of both seas is always of the same height, and therefore, because of the north-south orientation of the sea, this current is always flowing towards the south. The surface of the Baltic is much higher than the Gulf of Kiel; according to Nordenanker[1] 8' [feet] higher, according to Woltmann at Kiel 1'2" [feet, inches].[2] Observations made in the sailing area have revealed that the southwards current is two and a half times more fre-

1. *Von der Strömen der Ostsee*. Leipzig (1795), p. 8.
2. *Poggendorffs Annalen* II, p. 444.

quent than the northwards current.[1] The wind is thus blowing in the same direction as the current, from the south. However, only the north wind and the northwest wind, of which the former is rarer in our country, induces a northerly current, the other winds a southerly one. The winds thus also induce the southern direction of the current.[2] According to observations by Admiral Klint, the weight of the air pressure, indicated by a barometer, and not direct dependence on the direction of the wind, can cause a noticeable change in the height of the water surface (about 5' [feet]), and can thus also affect the direction and velocity of the current. The normal height of the surface is greatest in August and least in April. The reason for this relationship is obvious; for in the very months in which the water is highest, southerly currents are less frequent than northerly, and when a northerly prevents a quantity of

1. Schouw, *Skildring af Veirligets Tilstand i Danmark* (1826); p. 526 states that it is necessary to observe the current of the Sound in different places, as one should be aware that the current is different in the middle of the Sound and at the coasts; but it will also be necessary to examine the direction of the current at different depths; for it is experienced in general and almost daily by the fishermen, who lower the nets in the depths, that the current at a greater depth mostly is southerly, although on the surface it is northerly. Perhaps it can thus be established as a general law that in the Sound the current is always southerly; when the winds produce a northerly current, it is only on the surface.
2. *Den danske Lods* (1843), p. 104.

water from the Baltic from flowing, it is necessary that the surface in both this and the Sound becomes higher.

Periodic changes in surface elevation caused by the daily lunar movements around the Earth can be observed very little in the northern part of the Sound, which abuts the Skagerrak-Kattegat coast, and in the middle and lower parts not at all.

CHAPTER 2.

Geological conditions of the Sound.

§ 6. However, as the geological nature of the bottom in particular is to be investigated here, it will also be necessary to investigate the geological conditions of the coastal area, partly because the bottom must be regarded as a direct continuation of it, and partly because the questions about the time of formation of the Sound and what changes it has undergone after the formation must be clarified. We will now address these issues.

When and how was the Sound formed?

§ 7. Since the Sound intersects layers from the Tertiary period[1] it is impossible that it was formed before the end of this period. The famous Forchhammer believes that it was formed in a prehistoric time, when Bornholm's very ancient marine sediments (*havstokke*) had not yet formed and the Cimbrian flood[2] had not occurred, and he makes a very compelling attempt to explain how the Sound and other straits in the bays of Denmark have been

1. See Forchhammer, *Geognostisk Kort over Danmark* (1843).
2. See the text with comments.

formed. He believes that the Gulf of Bothnia was transformed into a lake after the land areas around it rose; a way was opened to that lake for a quantity of water flowing from the surrounding areas, through the less solid formations by which the southern part of Sweden was at that time connected with Russia. This excellent geologist now thinks that a force of violently turbulent water paved itself a way by breaking through in several places, and that the Sound was also formed by such a breakthrough.[1] To explain the formation of the Sound in this way, it certainly seems necessary to state that for the most part it was earlier, as it in fact consists of bays, one facing north, the other facing south, and that Zealand and Scania were connected by just one or maybe two narrow land bridges. Taking into account the shapes of these lands, by its nature with the coasts facing each other, and the distribution of islands and lowland areas, it seems to me very likely that these coasts, both at Helsingør and Helsingborg, and at Copenhagen and Malmö, were linked, and that between these land bridges was a lake. Assuming that the Sound was land all along its extent, it is not easy to understand how a single breakthrough could have occurred in one flood.

---

1. G. Forchhammer, *Skandinaviens geognostiske Forhold, et Foredrag holdt d. 22 Novbr.* (1843).

What changes did the Sound undergo after its formation?

§ 8. To ascertain the changes of the Sound, it is necessary to examine the nature of its coasts; we will, therefore, attempt to give a short overview in the following. Also in this case, there is a very large difference between the coasts; if we start from the southern coast of Zealand, which is adjacent to the Sound, we will find from the northern part of Stevns (from a place called Bøgeskov [Beech forest]) more or less to an inn called Flaskekro, many rows of marine sediments,[1] which together with the parallel coast extending one-eighth or one-fourth of a Danish mile to land, offers barren fields called "the Moor" or "the Heather". In the northern part of the coast the sediments consist of stones, in the other parts they consist of sand and seagrass [*Zostera*]; this difference seems to have arisen because the Swedish coast affords better protection against the attack of the waves. In a larger part of this section there is a sediment of exceedingly large size that protrudes above the others, like an isthmus, whose southern part, which simply consists of stones, appears so artificial that those living around think that it is artificially manufactured; but it is very evident that it was formed by a very violent storm. So, the coast of the southern part of the Sound was greatly increased during this period of the

1. Havstokke, Geschiebebänke.

Earth's history; it seems that the Sound was narrowed only by a sedimentation of the soil and not by land uplift.

The part of the coast of Zealand that surrounds the middle part of the Sound clearly shows that it is being swashed by the sea, which moves much slower than the sea on the southern and northern coasts; the sediments found here are very sparse. Moreover, the sediments and erosions vary according to the different nature of the coast; where it is flat there are the first type, where it is steep the others. Erosions occur at Vedbæk, between Sletten and Humlebæk and at Snekkersten, as evidenced by the fact that fishermen found tree trunks with roots with sand in the sea farther from the coast as it is today.

The force of the violently turbulent sea is especially evident throughout the entire northern coast, with the exception of a few protruding places, viz. at Hellebæk and Nakkehoved where the coast is steep, which has been increased by sediment consisting mainly of sand. Only at Hornbæk do sediments occur in many parallel rows, which, consisting of sand and stones, resemble the rows on the southern coast. The amount of the sand is particularly evident and remarkable at Villingebæk, because it is very similar to a phenomenon like land uplift. At a hill distant from the coast through which a path leads, one can see a layer of fossils (Fig. 2 b), which originates from the species now living in the sea, *Buccinum undatum*, *Neptunea antiqua*, *Littorina littorea*, *Cerastoderma edule*, *Arctica islandica* and others enclosed in a large amount of horizontal layers of sand, so that at first glance we seem to find the influence of the sea. But various animals that dwell at such different depths, such as *Cerastoderma edule* and *Arctica islandica*, and artificial things that are there, primarily a large amount of charcoal, and other similar things, easily convince us that this high is a sand dune called "klit" so that it shows how much the influence of the wind can mimic the influence of the sea. Something similar can be observed at the shores of the river flowing out at Villingebæk; at the bottom of the stream just one-eighth of a Danish mile from its outlet, a large amount of shells of marine molluscs, such as *Cerastoderma edule*, *Macoma calcarea*, *Tritia reticulata* and *Littorina littorea* are easily noticed. This case, which seems so unique, is solved by excavating the nearby areas; beneath a layer of sand, at a depth of one foot, is a layer of these marine organisms, sometimes only at a depth of half a foot, as the shells are exposed by a stream breaking through. This is easily seen in places that today are fields pervaded by a stream, which once was a sea bay, and which has been filled with sand in much the same way as Hanvejle and Bygholmvejle, two

Fig. 2.

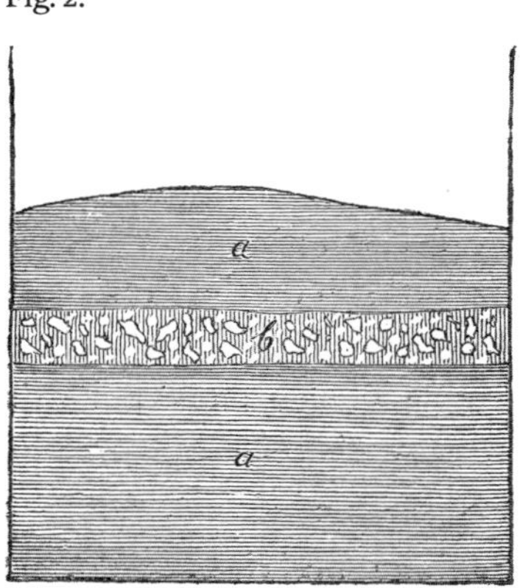

of the remote fjords of the Limfjord.[1] Lyngbye, blessed in remembrance, is said to have found marine animals under the same conditions at Søborg; from this it becomes very likely that a bay of the sea has also penetrated straight to Søborg, so there can be no doubt that the lake, called Arresø, was a bay of the sea. The amount of moving sand has been more violent here.

We therefore understand that the changes that the coast of Zealand has undergone continue today, as they are the results of gathering of sediment and eroded sediments. If more violent changes may have taken place, a minor land uplift must necessarily have occurred; whether it was a great thickness of marine sediments over the sea caused by violent storms, or if land uplift caused it, I dare not decide.

A part of the coast of Scania, which surrounds the Sound, clearly shows that a much greater area has been changed by erosion than by sedimentation. It is generally said[2] that this applies to the coast extending from Falsterbo to Landskrona, which I have not, however, examined myself, but that the other part has been eroded, I have experienced myself. This has especially been the case with the steep coast, which extends very far from Landskrona to the north beyond Glumslöf, where it appears that there was once a lake near the sea; on the very steep peak there are both sev-

1. Forchhammer, *Studien am Meeresufer* in Leonhards & Bronns, *Neue Jahrb. Für Miner u. Geogr.* (1841), p. 10.

2. Cf. Cateau-Catteville, *Beskrivelse af Østersøen*, translated by Rawert.

eral chalk layers formed in freshwater, and then to the north a layer of fossil peat (Tørv). Within the steep layer there are lower places, quite reminiscent of desiccated lakes, and even in the steep area, channels through which water could flow into the Sound. A little lower at

Fig. 3.

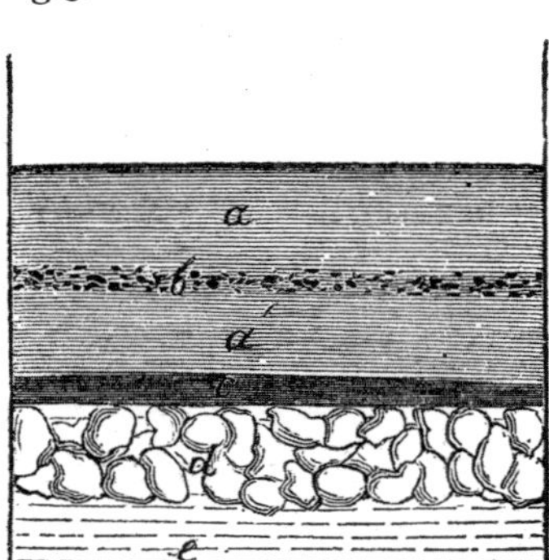

Frederiksleje, near the coast, about 15' [feet] above sea level a ditch was dug in which these layers were exposed (Fig. 3): above a layer of soil mixed with sand, one and a half feet thick (a), then a thin layer of marine molluscs, such as, for example, *Littorina littorea*, *Hydrobia ulvae* and others (b), then a layer of the same type as the upper one (a'), then a layer of pebbles (d), under which clay (c) is found. Whether this formation should be considered to have arisen from a flood whose surface was higher than fifteen feet, or land uplift, I dare not decide, but I consider the first explanation the most probable. Therefore, there seems to be only evidence of a rise of the coast of Scania during this period, but I have not been able to find any signs of a subsidence of land at all. In summary, we have found changes that the coasts around the Sound have undergone after the formation of the land during this period.

1) Neither the coasts of Scania nor Zealand have been altered by land subsidence.
2) These things that appear to be signs of land uplift are on both coasts very small and of such a kind that it is difficult to determine whether or not they can be ascribed to sediments with the same confidence.
3) The entire coast of Zealand, except for a few places where it is very steep, has increased by the sedimentation of soil, which is primarily in line with the southern and even more so with the northern coast.
4) Nearly the entire part of the coast of Scania, which is adjacent to the Sound, is constantly diminished by erosion; how much is gradually removed from it is not easy to determine.
5) After the Sound was formed, it seems to have changed only by sedimentation and erosion of soil, not by land uplift or subsidence.

NOTE. Before leaving the description of these shores, we will look more closely at the whole nature of the clayey slopes, though not necessarily related to the entire description. On the coast of Zealand, there are two places on which the clayey slopes are so steep that all layers are exposed, at Vedbæk and between Sletten and Humlebæk. The slope consists mainly of plain yellow clay, only there is a layer of larger erratic blocks of rocks at the top, but scattered all around, and in the middle there is a layer of much softer yellow clay. A larger part of the upper slope (Fig. 4) consists of plain clay (aa), but here there are three denser layers of scattered larger and smaller erratic blocks (ccc), separated from each other and from the other clay by thinner layers of softer clay. At the bottom of the slope, here and there blue clay occurs, and in one place a layer of compacted lime (Calcareus conglomeratus) is so large that it is transported to Copenhagen by ship and used instead of cement. It extends slightly into the sea as the loose clay with which it was covered has been eroded. On the Swedish coast at Glumslöf there is a slope that is longer and higher than the others on the coast of Zealand (Fig. 5). There is a rather large difference between its various parts. However, this should only be stated in general terms. About half of the upper part consists of plain sandy yellow clay, along with many erratic blocks of rock. Above, there are two or three alternating layers of lime containing fresh water (ccc) and soil (bbb), but the lower half consists of clay that can be split in layers, which is blue, plastic and completely devoid of rocks. However, many curved layers of blue clay should be noted (d), which correspond completely to the curved layers (cauliflower-bent) discovered by Forchhammer.[1] There can hardly be any doubt that this is the same plastic clay that can be split in layers, as Forchhammer has described, and which he believes has produced a large formation where the southern part of the Skagerrak-Kattegat bay is today.[2] This is the outermost boundary we know of this formation to the south.

Fig. 4.

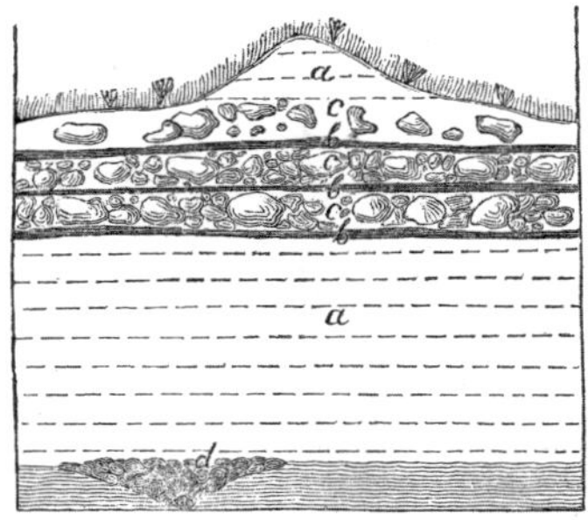

Fig. 5.

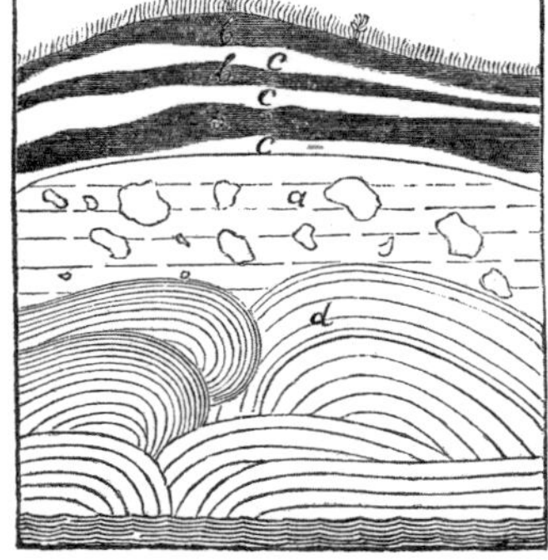

1. H. Krøyers *Tidsskrift*, vol. 1, p. 209.
2. Ibid., p. 216.

About the nature of the bottom of the Sound.

§ 9. **Description of the bottom.** Above we have seen that the sound is a rift valley or rather a very even gorge, which for the most part intersects Tertiary layers, sand and clay. We will now examine whether these layers are connected in the same way through the bottom of the Sound as on land. We will easily find that it is not so, but that the dissolved masses of the bottom are formed by their own law, so that there exist three forms of bottom, namely 1) one consisting of clay, 2) one consisting of stones or seashells, and 3) one consisting of sand. An area with clay of very different width and distance from the coast dominates the middle part of the Sound.[1] In the northern part of the Sound, at Kullen, it extends all the way towards land, from where, all along the coast of Scania, it is not far from Kullen to Landskrona; in the Danish part there is no bottom of compact clay near the coasts except between Hellebæk and Snekkersten, at a distance of scarcely one quarter of a Danish mile from the coast. In the rest of the Sound there is hardly any clay bottom except at a distance of one or two Danish miles from the coast. Close to all the coasts of the island of Hven, which rises in the middle of the Sound, there is a bottom of clay. This area is only interrupted between Helsingør and

1. Indicated with yellow on the plate.

Helsingborg, where there is a sandy bottom. On both sides of this zone is a very narrow zone, a rocky area,[1] where the bottom is covered with larger or smaller stones. From Kullen to Landskrona it is limited by distinct boundaries. In the rest of the Sound, the stones are more scattered, so that this area almost disappears. At Zealand, it only exists at a small distance to the south at Helsingør off Snekkersten and Espergærde, and along the western coast of Hven surrounded by distinct boundaries. In some places, instead of stones, there is a large amount of shells, first and foremost of ocean quahog (*Arctica islandica*), in the same way as off Hellebæk. The fishermen call the narrow space where the bottom consists of accumulated shells *Skallekant* [the shell's edge].

Within the stony zone close to the shore a sandy area exists.[2] The bottom consists of sand and stone, first and foremost of sand, although stone is predominant in some places. It extends over the greater part of the area, extending towards the south from Helsingør–Helsingborg, and also over the southern part of the Sound. Only off Kullen is this area missing because it lies on a clayey coast. From Helsingør, where it is very narrow, it constantly grows in width right to Køge, where it covers the entire area between this town and Falsterbo. The same is true in the Swedish part from Helsingborg to Falsterbo. It exists around Hven, though it is very narrow.

1. Indicated with dark brown and red on the plate.
2. Indicated by light brown, green, blue and white on the plate.

Explanation of the formation of the three geological areas.

§10. A definite and delimited relationship, which always shows itself between the natural properties and depth of the bottom, easily shows the laws by which it is formed. So, we see that the sandy bottom extends from the shore to a certain depth, 7–8 fathoms, then it gradually becomes more clayey, so that in almost in all places of greater depth there is a bottom of clay. Here, that is, closer to the coast of Scania than Zealand, one finds that the depth is more similar to the first than the last, as we saw above. If there were no other movement of the sea than that of the waves, then the dissolved constituents would have been scattered so that the number of particles inversely corresponded to the depth, so that the very soft clay would be found at the greatest depth and the largest stones closest to the coasts. But we will see that this relationship is influenced by another movement of the sea, namely that which is caused by the current; that was the cause of the third stony area. It is very easy to observe that this is particularly the case where the power of the current is more violent because of the coastal morphology, and is therefore induced by the erosion of sand and clay, leaving remaining stones alone. Therefore, a strong current in the very narrow parts of the Sound between Helsingør and Helsingborg results in the absence of dissolved clay in the depth.

It must therefore be stated that when the water first broke into the narrow gorge, which at that time separated Zealand from Sweden, the layers at the bottom of the narrow gorge were divided in the same way as we find today on the coasts that surround the Sound; so, sand and clay have been mixed with each other. But now they are divided differently because of two counteracting conditions – the force of the moving water (the basis of change and movement) and the depth (the basis of damping and resisting change). By this time there was soft clay and very small particles of sand in all the places, from the shore to a depth of 7–8 fathoms, where the depth was shallow enough that the movement of the waves also affected the bottom, and they have been transported by the water, and then moved towards those parts of the Sound whose depth was so great that the water movements could not affect them, i.e. in places deeper than 7–8 fathoms. Thus, it is easy to understand that an area of clay that fills the very deep parts of the Sound is surrounded on both sides by an area of sand. Likewise, everywhere where the bottom rises, sand necessarily appears. Thus, all shallow areas consist of sand and are like islands surrounded by a zone of clay. We can therefore also say that in the Sound, all the higher-lying areas, the shallow areas, i.e. the sea heights, consist of clay [we believe that this is an error and should be sand]. This assertion is corroborated by a long-stretched shallow area that faces south from Helsingør and is called the Disk [Disken].

It is clear that the power and depth of the movement must be examined if the formation of the western coast of Schleswig, known as marsh, is to be examined. It shows that clay can be sedimented at shallow depth if the movement is slow (which occurs along the range of islands off the western coast of Schleswig), but at a greater depth sand can be sedimented if the movement of the sea is more violent (which it is off these islands).

Thus, based on this section, we have concluded the following.

1) If the depth of the sea consists of loose sediments, it consists of three kinds: 1) sand, 2) clay, 3) stones or aggregations of shells.
2) As the nature of the bottom appears to be one of these three kinds in the sea, it depends on the relationship of depth to the forces of the waves and the current.
3) Where the depth is not so great that the bottom cannot be affected by the movements of the waves (in the Sound, from the coast to a depth of 7–8 fathoms), sand appears, with the clay (whose particles are much smaller) washed off.
4) Where the depth is so great that the bottom cannot be affected by the movements of the waves (in the Sound, from a depth of 7–8 fathoms), clay, which is washed off at places in the sea of less depth, and sedimented here.
5) A bottom consisting of sand thus indicates shores and lower-lying places with clay indicate depth.

6) A bottom of rocks or shells indicates areas in the sea where the current has been very violent.
7) This distribution of substance should only be regarded as a phenomenon at the surface of the bottom.

The results from the previous section transferred to the geology.

§ 11. I do not think that it has so far been sufficiently understood how important a study of the nature of the bottom is to illuminate the geology, although it will be very clear, partly if we would consider that most geological formations that are called normal have emerged at the bottom of the sea, partly if we would remember that the view of those who have the opinion that the ancient formations can be brought into line with those that existed during this period of the earth, is constantly being confirmed ever more strongly. However, although knowledge of how the sea affects the coasts has been greatly increased by the famous Forchhammer's studies),[1] we will in the following, first and foremost, if we want to compare how the sea at different depths affects the bottom, see that it becomes clear how important these studies are. In the foregoing, we sought to demonstrate the unity within the diversity of the sea floor, showing that there are but three principal types of bottom. And we may further prove that

1. Leonhard & Brenn, *Neues Jahrbuch für Mineralogie und Geog.* (1841), p. 10.

the causes of these variations, throughout the successive periods of the Earth—or, what is the same, within the strata of the normal formations—exhibit the same underlying unity, allowing us to surmise that the same or analogous forces have prevailed throughout.

Depth of the Sound compared to Denmark's Tertiary formation.

§ 12. Geological formations that belong to the Tertiary system consist of clay and sand and cover almost the entire area of Denmark. They are divided into three parts by the famous Forchhammer: 1) a formation of electricarbo (Amber brown coal formation), 2) of Tertiary [glacifluvial] clay (Rullestensleer), 3) of Tertiary [glacifluvial] sand (Rullestenssand), which he believes has formed both in different times and in different ways.[1] By examining the depth of the Sound, however, a certain doubt has arisen for me as to whether the last two parts can be attributed to a different time of formation, and whether they have arisen in the way he thinks. I would like to admit that it is not without some sense of shame that I dare to convey opinions that diverge from this excellent geologist. Just as I am ready to withdraw my view if it is disproved, so I very much acknowledge that it is due to this highly educated man, according to whose instructions I have organized my investigations, if I have the correct opinion

1. See Krøyers *Tidsskrift*, vol. 3, p. 546.

of anything concerning these subjects. If we see how these two parts of the formation are distributed, we find sand in the highest places and in all the great heights and only clay in the deeper areas.[1] This diverse distribution allows me to clearly indicate diverse causes of how they are formed. If the bottom of the Sound were [imagined to be] raised above sea level, we would see a smaller picture of the conditions that the two parts of the Tertiary formation present on a larger scale. All higher places that were previously the low-lying places of the sea consist of sand, all valley areas that were previously deep areas consist of clay. By comparing these conditions with Denmark's Tertiary formation, which has certainly similarly formed the seabed that previously covered Denmark, we readily realize that it is necessary that the part of the formation, called Tertiary sand, today forms hills, the low places of the sea, and the part of the clay called the plain has been the greatest depths of the sea; thus, these formations did not occur at different times, but at the same time under different conditions in the same sea.

Also, the formation of Denmark's areas, which are equally low and consist of Tertiary clay, and of those which merely consist of sand, can be explained from the previous; one must assume that they are formed under the same conditions under which the narrowest part of the Sound is formed, where we also find sand at a depth that otherwise exhibits clay. (See the plate on geographical conditions between Helsingør and Helsingborg).

---

1. See Forchhammer, *Geognostisk Kort over Danmark* (1843).

NOTE 1. The truth of this view concerning the Tertiary formation of Denmark may perhaps be objected to on the grounds that these two portions contain different fossils—or rather, that one, the clay, is entirely devoid of fossils, while the other, the sand, contains the remains of animals which still inhabit our seas today;[1] from this it is certainly possible to conclude that these formations appeared at different times. But first it should be noted that although fossils have not yet been found in the clay, they may be found later, which is more likely because we can search a very large part of the sand without finding fossils, because they are very rare in this formation. However, if it can be observed that it is certain either that different fossils are contained within them, or that fossils are merely in another formation alone, I believe, however, that these formations could have formed during the same time. Also, in the present sea, the animals of the sandy bottom are very different from the animals of the clay bottom, as we will show below. Perhaps it will also be objected, as I myself have previously noted, that this distribution of soluble matter only occurs on the sediment surface (see § 10), that is, the low-lying places of the sea, and also the mounds of the earth, which belong to Tertiary sand, if there is a real correspondence, alone in the surface consists of sand, there is always a core of clay. To this I respond that, until the contrary is proven, I believe that this explanation fits equally well in the deeper areas of the sea and the shallower ones. Finally, it should be noted that if there are problems in my explanation that I cannot clarify now, then I do not understand how it is not the case that analogous explanations necessarily appear between the nature of the present bottom and the previous one.

NOTE 2. There are other views of the same excellent author on the geology of Denmark that I cannot fit together with the results I have obtained by observing the effect of the sea in the depths. The islands that occur in

---

1. See Krøyers *Tidsskrift*, vol. 3, p. 548.

the southern part of the Sound to the west are considered to be of clay, because a flood, which is believed to have flowed from the east and flooded the whole country, eroded the sand.[1] From the previous, it is readily understood that a violently turbulent sea cannot erode sand unless simultaneously the clay is also eroded, because the latter consists of much smaller particles that more easily are eroded by the movements of the sea. Therefore, it seems to us that such a flood rather had the effect of washing the clay away while leaving the sand behind.

As this very famous geologist has expressed elsewhere,[2] his opinion is that there is a layer of clay between the layer of green sand and the limestone layer on the island of Saltholm. It is therefore confirmed that clay exists north of Saltholm at a depth of seventy-two feet. From the previous, we know that wherever clay is found at this depth, at least only on the surface, no safe conclusion can be drawn about the nature of the layers at a greater depth.

The relationship between the nature of the seabed and the general geology.

§ 13. Although it has been previously stated that the rocks in all formations called normal can be attributed to one of three kinds – 1) schist, 2) layered sandstone, 3) limestone – the internal correspondence between these three parts and their relation to the whole analogous parts of the dissolved masses that the seas have today even during this period of the earth have so far hardly been sufficiently noticed. We see that the sea, in accordance with con-

1. *Skandinaviens geognostiske Natur*, a speech given on 22 November 1843 by Forchhammer, p. 18.
2. Forchhammer, *Danmarks geognostiske Forhold*, Copenhagen (1835), p. 49.

ditions determined by its movement, depth and current, deposits sand or clay or lime. We will now endeavor to show the analogous parts of the formations recognized by all.

The nature of any normal formation is raised soil of original rocks. Their normal formations, i.e. those that, in layers deposited in the sea, contain residues of organisms, always appear between certain boundaries of the surrounding sea of masses, which are separated by violent and frequent rises and depressions. Since the original substance has always been the same, namely granite, we can see clearly why the rocks of all formations can be attributed to analogous parts. We know,[1] in particular, that clay and sand are formed from eroded granite and gneiss, and that at the same time lime has emerged from the interior of the earth. As to this, it is of less importance whether these rocks are produced by schist [Schistus argillaceus] or dissolved clay, or by sandstone [Arenarius Schistus] or dissolved sand, or whether it is limestone or chalk [Calcareus or Crete]; therefore, only the modifications of approximately the same substance, as a result of external influences, are indicated. The following overview shows the analogous parts of parallel series which form the basis of different formations, so that also in geology the view that is valid in the systematic distribution of animals and plants is confirmed, that all development in nature takes place through parallel series of analog parts.[2]

---

1. Forchhammer *om Leerarternes Oprindelse in Vidensk.* Selskabs Skrifter (1832).
2. See *Havetidenden* volume 9, issues 5 and 11, and *Entwurf einer systematischen Eint. Der Platwürmer von Ørsted* (1844), p. 33.

Overview of analogous parts of all formations.

I. Silurian formation.		
1. Clayey sand (Graywacke)	2. Schist	3. Transitory limestone
II. Carboniferous formation.		
1. Carboniferous sandstone	2. Carboniferous schist	3. Carboniferous limestone.
III. Red sandstone formation.		
1. *Red sandstone*	2. Copper-bearing shale	3. Alpine limestone (Zechstein age)
IV. Variegated sand formations.		
1. Variegated sand	2. Variegated clay (marl) (Keuper age)	3. Shelly limestone
V. Jurassic formation.		
1. Sand of Lias age (= Black Jurassic).	2. Shale of Lias age.	3. Jurassic limestone
VI. Cretaceous formation.		
1. Green sand	2. Sand?	3. Chalk
VII. Tertiary formation.		
1. Tertiary sand	2. Tertiary clay	3. Limestone?

In these series the formations enumerated in the 1st correspond to sand, those in the 2nd to clay, and those in the 3rd to accumulated shells of molluscs, and if the analogy is correct, they all emerged in the same way.

These three parts are not equally prevalent in all formations; in the oldest, clay is very prominent, as clayey shale is much more prevalent than in the other formations, but there is also much clay in the part that corresponds to sand, because graywacke is sand interlaced with clay. Next, sand formations begin to be more frequent until one reaches the final stage of development with the formation called variegated sand, so that the clay-like part diminishes and appears less clearly in the Keuperian formation as it stands just for the most part that consists of clay or marl. After this formation, the third formation, limestone, is most prominent until the Tertiary formation. All formations older than the Tertiary can therefore be attributed to three sections, depending on whether sand, clay or lime is predominant.

1 *Clay section.*
  Graywacke formation.
2 *Sandy formations.*
  Carboniferous formation; red sand formation; variegated sand formation.
3 *Calcareous section.*
  Jurassic formation, Cretaceous formation.

At the time when graywacke was raised, there were no smooth shores that could produce sand. A deeper sea surrounded the islands, so that the bottom, which was raised, consisted almost of clay alone. The formations that then follow directly to Jurassic formation exhibit large coastal plains. The famous Forchhammer has already proven[1] that the formations of sands are closely linked in both Scandinavia and Bornholm. These coasts were covered with forests, which, after a subsidence of the sea, had been filled with sand and turned into coal. The same famous geologist has proven[2] that these are phenomena that are exactly the same as those that occurred at a much later date on the west coast of Jutland; but here the forests form petrified land, and the sand has not changed to sandstone. Likewise, this period's frequent formations of fossil salt exhibit phenomena very similar to those now occurring on the salt plains of southern Russia.

The conditions during the limestone period must necessarily have been such that they were primarily for the benefit of calcareous animals; namely, layers consisting of their compressed remnants have a much greater extent here than clay and sand layers. From this it can be understood that mechanical forces that had the greatest influence in the past have been overcome by organic forces.

1. Falck, *Statsbürgerliches Magazin. Videnskabernes Selskabs Skrifter* 7(7): 19.
2. *Oversigt over Videnskabernes Selskabs Forhandlinger i 1842*, p. 64.

CHAPTER 3.

About the areas with algae in the Sound.

§ 14. It seems peculiar to everyone who, like me, begins a study of algae in the Sound, that the writings about the northern algae show that the authors have too much left the sea that was closest to them almost out of consideration. Lyngbye's *Hydrophytologia Danica*, which deals with the Danish seas, relies primarily on studies of the Fjord of Odense, and also in the famous Agardh's writings on algae, the distribution of algae in this sea is particularly highlighted, so that general laws are based on these observations. In this case, Hornemann follows the others. In recent times, however, more weighty aids have been used by J. Agardh[1] and Liebmann.[2] In particular, the first-mentioned author's investigations are of great importance because from them, the laws of the general vertical distribution appear. In short, they are as follows:

"According to the geographical distribution, the Scandinavian algae fall into three zones corresponding to three subdivisions of algae:

1. *Novitae florae Sueciae ex Algarum familia*. Lund (1835).
2. Krøyers, *Tidsskrift*, vol. 2, p. 464.

1) The zone of green algae encompasses the entire freshwater vegetation and, to a greater extent, amphibious sea plants, which is why the central area of this region, if you look at thalassiophytes only, is found on our eastern shores. This area can be divided into two:
   (a) The area comprising freshwater algae.
   (b) The area of the green algae. In this area, *Ulva*, *Chaetomorpha linum* and *Cladophora rupestris* dominate.
2) The zone of brown algae, in a way a middle ground between green and red algae. Brown algae prefer more saline seas, possibly preferring more open shores, though they do not avoid less salty seas, but occur in more constricted forms in them. They may be found in the following regions.
   (a) The area of lichens.
   (b) The area of *Sphacellaria* in the lower water trenches [rock pools], often exposed towards the sea.
   c) The area of the *Fucus* at the border [shoreline] of the sea itself.
   d) The area of *Dictyota* in the basins at 3–6 fathoms below the sea surface.
   (e) The area of the *Chordaria* on rocks, which is particularly exposed to the waves of the sea.
3) The zone of red algae in the open sea, at a depth of about 6–14 fathoms.
   (a) The *Chondrus* area at depths that are not too exposed to the sea.
   (b) The area of *Delesseria* at a depth of 9–10 fathoms."

Although, as one can understand from the following, we agree with the famous algae scientist in the essentials that algae can be divided into three zones corresponding to three areas of their distribution, we believe that the horizontal and vertical distribution is not sufficiently clarified in this author's subdivision. I have not observed it in the study of the distribution of plants in the Sound, which I have previously published.[1] It should be noted that the distribution of algae from the coast to the depth corresponds to the distribution of plants from the top to the lowest part of a mountain, and therefore indicates a vertical direction. Therefore, the divisions should be called areas and not regions as with this author. Therefore, the distribution of algae in the Sound in the following is studied both vertically and horizontally. This author also lacks observations on how the distribution is in the deeper areas of the sea.

§15. As we move from the Zealandian to the Scanian coast, we will see in the depths of the Sound what different areas of plants occur sequentially on both sides of the enclosed depth that we have to pass over. Initially, *Oscillatoria*, *Conferva* and *Ulva*, all green algae, are found. Below them at a depth of 2–5 fathoms, fucoid brown algae begin to follow and at a slightly greater depth, laminarias, i.e. all brown algae characterized by an olive color. At a depth of 8–10 fathoms, the brown algae begin to

1. *Forhandlinger vid de Skandinaviske Naturforskarnes tredje Møte i Stokholm* (1842), p. 621.

slowly disappear, and the red algae (*Chondrus*, *Delesseria*, *Ceramium*) follow. It is thus these plants that are found at the greatest depth, but in the middle of the Sound a zone devoid of any vegetation extends through almost all of the area described above with the name clay zone. These three areas occur on both sides of the Sound, but in different places of varying extents. These differences are easily noticed by studying the tables, one of which (Plate I) outlines the areas viewed from above, the other (Plate II) viewed vertically. The coherence of the plant areas found on the rocks can be considered a reason to call them [algal] areas. Now we will examine the individual areas in a little more detail.

The area with green algae.

(On the plate they are marked in blue and green).

§16. This area, which extends from the coast itself to a depth of 2–5 fathoms, is characterized by the green algae, *Oscillatoria*, *Conferva* and *Ulva*. They are not entirely lacking anywhere in the Sound, but are of very different prevalence. In the southern part of the Sound, the distribution is most extensive and gradually smaller further north. This is due to the fact that the southern part is of approximately the same nature, and there are lakes where green algae particularly dominate. This area can be divided into two sub-areas: 1) a sub-area with *Oscillatoria* or blue-green algae [Cyanobacteria], 2) a sub-area with Ulvacea.

The sub-area with *Oscillatoria* or blue-green algae. This sub-area is the part of the area with green algae, which lies up the coast and which very often is completely devoid of water. Studies previously conducted by me at Hofmansgave have already shown that blue-green algae dominate on the flat beach very close to land.[1] In shallow water in the Sound, blue-green algae also occur to a large extent among all the algae. Thus, there occur *Blennothrix glutinosa, Lyngbyea aestuarii, Spirulina subsalsa, Microcoleus chthonoplastes* and *M. fuscus* in shallow areas with a low salinity, i.e. in stagnant areas close to the shore without any connection to the sea after the tide has fallen. The plants in these places must be mentioned separately and they are: *Cladophora fracta, Chaetomorpha ligustica, Chaetomorpha linum, Ulothrix floccosa, Ulva clathrata* and *Merismopedia litoralis*. On the rocks at the fortress Trekroner at the shoreline, one easily encounters *Lyngbyea lutescens* and at Kullaberg *Calothrix fasciculata*. They are found so frequently on the rocks that when the water recedes, they appear to be covered with black color.

The sub-area with Ulvaceae. This sub-area covers a part of the area with green algae that only rarely or never lies above the surface. One can easily understand that here in such divisions no fixed limits can be stated; often some plants can be attributed to different subdivisions with equal justification. Of the green algae that occur at the sur-

---

1. Krøyers *Tidsskrift*, vol. 3, p. 552. *Beretning om en Excursion til Trindelen i Odensefjord* by Ørsted.

face of the sea to such an extent that there can be doubt that they are to be attributed to the former sub-area, mention must be made of: *Ulva intestinalis, U. compressa, U. linza, Chaetomorpha linum, Spongomorpha aeruginosa, S. arcta, Urospora penicilliformis, H. assimilis* (= *Urospora penicilliformis*), *Ulothrix contorta, Cruoria pellita* and others. Particular to this sub-area are: *Ulva lactuca* and the form *latissima*, which coasts with muddy bottoms in particular contain in large quantities, such as Kalvebod Strand between Copenhagen and Trekroner, at Landskrona and other places. Meanwhile, they appear at a depth of 3–5 fathoms. *Cladophora rupestris, C. hutchinsiae, C. glomerata, C. flexuosa* and *C. elegans* (= *C. flexuosa*) occur at a depth of 6–7 fathoms.

NOTE. Although green algae dominate this area to such an extent that they can be said to constitute the entirety of the vegetation, there are, however, some brown and red algae; but I cannot name any red alga from this area whose color is beautiful. Of these, the following should be noted: *Lichina confinis, Petalonia fascia, Scytosiphon lomentaria, Pylaiella littoralis, Ectocarpus siliculosus, Spongonema tomentosum, Ceramium diaphanum* and *Vertebrata fucoides*.

List of green algae.

§ 17. Green algae.

Acrosiphonia arcta at Kullaberg. *Elachista globulosa* in the stagnant sea water on the island Gråen by Landskrona Ag. – *Spongomorpha aeruginosa* Kullen-

Trekroner. – *Cladophora sericea* Kullen-Trekroner – *Lychaete battersii*. – *Cladophora glomerata* – *Cladophora rupestris* – *Acrosiphonia arcta* Dillv. Kullen-Trekroner. *Spongomorpha arcta* Trekroner. – *Cladophora flexuosa* Kullen-Hven. – *Cladophora flexuosa* between Taarbæk and Hven[1] – *Cladophora hutchinsiae* [2] (Cf. Hutchinsia Flora Danica. vol. 2, 314) between Copenhagen and Trekroner. – *Conferva obtusangula* Lgb Taarbæk. – *Rhizoclonium riparium* Helsingør. Liebm. Trekroner. *Chaetomorpha linum* – *Chaetomorpha ligustica* (*Zygnema littoreum* Lgb.) Near Copenhagen (Ny Badehus) – *Chaetomorpha linum* near Landskrona by Gråen Ag. – *Chaetomorpha melagonium* at Helsingør Liebm. – *Chaetomorpha linum* at Hellebæk. – *Conferva Hofmanni* Ag. Trekroner Ag. *Tribonema bombycinum* (*Conferva bombycina submarina*?) Near Copenhagen (Kalvebodstrand by Ny Badehus[3]). – *Urospora penicilliformis* Near Copenhagen (on the rocks at Langelinie). – *Urospora penicilliformis* (*Conferva hormoides* Lgb). Kullen,

---

1. This very elegant species differs from the *Conferva fracta* (= *Cladophora fracta*) that it most resembles, in that the lower part of the joints grows out at the branches.
2. In the list of errata, mentioned as an error for *Conferva diffusa* Dillv.? (= *Cladophora hutchinsiae*).
3. Different from *Conf. bombycina* (*Bombycina stagnalis* nob.) (= *Tribonema bombycinum*) by half as short interstices between the joints and small, very closely connected knees. These two species form a very distinct genus, which in particular differs from all others in the nature of the spore mass. It can be distinguished in this way: single undulating filaments, slimy, simple knees more or less tied together so that they appear as simple transparent filaments; a single joint's secondary spores form two separate masses, attached to the knee and simply attached together by very thin filaments.

Helsingør Trekroner.[1] – *Ulothrix flacca* (*Myxonema flaccum* Liebm.[2]) Krøyers *Tidsskrift*, vol. 2, p. 323). *Ulothrix flacca* at Hornbæk and Helsingør. Liebm. – *Erythrotrichia carnea* (*Conferva ceramicola* Lgb., *Callithamnium ceramicola* Suhr., *Ceramium ceramicola* Ag.) on *Ceramium virgatum* near

---

1. A very important characteristic of the genus *Hormiscia* has, in my opinion, hitherto been completely overlooked; it consists in the fact that the spore mass is surrounded by a transparent zone, so that there appears to be a double tube, an inner and an outer one. In this way, an unusual size of the transparent part of the thread emerges. It is probably this feature that the famous Fries has referred to with the words: "The joints swell into soft jelly-like threads." The first three species can be distinguished in this way:

   Urospora penicilliformis (as *Hormiscia assimilis*). With cells that are about twice as long as they are wide, with knees that are slightly tied together.
   Urospora penicilliformis (as *Hormiscia penicilliformis*). With cells that are approximately equal in length and width, with knees that are very close together.
   Ulothrix flacca (as *Hormiscia flacca*). With cells that are twice as long as they are wide, with knees that are slightly tied together.
   In addition, they differ from each other by the different size of the inner transparent tube.

2. Plants from very different genera are associated with the genus *Myxonema* Fr. If one considers *Conferva lubrica* (= *Stigeoclonium lubricum* (Dillwyn) Kützing, 1845) as the type of the genus, all the other species are lobed and should not contain more zoospores in any joint. Of these, *Ulothrix flacca*, which with the characteristics of the genus is consistent with *Hormiscia*, can be completely associated with this genus. I wonder if *Myxonema curvatum* Liebm. is a species of *Scytonema*? *Conferva lubrica* forms a very distinct genus both in its branching and especially in the individual zoospores in each individual part; between *Confervae ramosae* there is an analogy to the genus *Ulothrix* among the unbranched forms.

Hofmansgave. – *Ulothrix subsalsa nob.* Near Copenhagen (on Kalvebodstrand by Ny Badehus[1]). – *Ulothrix* (*Bispora*) *floccosa nob.* (Cf. *floccosa* Ag.) Near Copenhagen on Kalvebodstrand, by the *Zostera marina* (common eelgrass[2]).

Bryopsidales.

Bryopsis plumosa Between Tårbæk and Hven.

Ulvaceae.

Ulva lactuca – *U. lactuca f. latissima* L. – *Ulva linza* – *Ulva compressa* – *Ulva clathrata* – *Ulva flexuosa subsp. paradoxa* (*Solenia clathrate confervoidea* Ag.). *Merismopedia litoralis*, as *Erythroconis littoralis* Ørsd. Krøyers *Tidsskrift*, vol. 3, table 7, ff. 1–3.

Oscillatoriales.

Rivularia atra Roth. – *Rivularia pellucida* Ag. At Sundets havplanter Ag., Unknown to me. – *Rivularia nitida* Ag. at Trekroner. – *Isactis plana* Harvey in Hooker's *British Flora*, vol. 2, p. 394 at Hven. – *Calothrix scopulorum* Ag. in the

---

1. Distinguished by threads that are flexible and twisted and form a very small tuft, by cells that are equal in diameter or one and a half times as long, in the middle a single zoospore with a single oval joint, surrounded by two colorless triangular spots.
2. This subgenus is characterized by a spore mass delimited by two zoospores; in the true Ulothrices there is only one zoospore in the middle joint. *Antithamnionella floccosa* is, by the famous Agardh, restricted to fresh water; but according to samples that Hofman Bang has sent me, it has also been found in the sea by this very knowledgeable algologist.

Ceramium at Kullaberg. – *Tabularia fasciculata* on the rocks at Kullaberg. – *Blennothrix glutinosa* (*Oscillatoria maiuscula* Lyb.) *Lyngbya aestuarii* Liebm. (*Lyngbyea ferruginea* Ag. *Oscillatoria aestuarii* Lgb.) – *Heteroscytonema crispum* Amager. *Lyngbyea lutescens* Liebm. Trekroner. – *Microleus fuscus* Ørsd. near Copenhagen (at Ny Badehus). – *Spirulina subsalsa* Ørsd.[1] together with the previous one. – *Lyngbya lutea* on the rocks by the beach near the town of Helsingborg on the very border of the sea Ag. near Copenhagen. – *Leucothrix mucor nob.*[2] Covers plants in stagnant submarine sea areas with its radiating threads, in the same way as slime.

Bacillariophyceae.

Melosira lineata (*Fragillaria lineata* Lgb.) at Trekroner, Copenhagen (Ny Badehus). *Gallionella moniliformis* Ehrb. (*Fragillaria numuloides* Lgb.) In the same place. – *Agonium*

1. Krøyers *Tidsskrift*, vol. 3, p. 566.
2. Characters for the genus. White filaments forming branches and cover layers with very long elongated rods, distinct cells, very simple hyaline interstitiae and without constrictions, numerous white powdery spores. This genus differs greatly from *Callothrix*, to which it is most closely related, both in the nature of the tuft that radiates extensively, and in having much more distinct interstices, as well as in the characteristic nature of its zoospores. Meanwhile, small movements can be observed as with the other Oscillatorineae.

centrale nob. At Trekroner.[1] – *Ceratoneis closterium* Ehrb. (Act. Acad. Berol. 1840). – *Navicula baltica* Ehrb. – *Navicula sigma* Ehrb. – *Navicula lanceolata* Ehrb. – *Navicula bifrons* Ehrb. – *Sigmatella nitzschii* Rütz. – *Amphiprora constricta* Ehrb. Act. Acad. Berol. 1841 pp. 410. – *Eunotia turgida* Ehrb. – *Navicula westermannii* – *Himantidium faba* – *Coconneis scutellum* Ehrb. – *Surirella tenuis*[2]. Is it a particular species? – *Homalodiscus vulgaris* nob. Very common among most marine plants.[3] – *Homalodiscus ovalis* nob. With the previous one. – *Diatoma vulgaris* – *Bacillaria elongata* Ehrb. – *Bacillaria flocculosa* Ehrb. *Diatoma marinum* Lgb. *Bacillaria adriatica*. Hyac. v. Lobarczuski Linnea V. 14. *Diatoma signata*, *D. sera* and *D. caelata*.[4] *Grammatophora mexicana*, *G. stricta* and *G. oceanica* Erb.[5] *Striatella unipunctata* Ag. – *Odentella aurita* Ag. – *Tessella catena* Ehrb.

---

1. Characters for the genus. Very thin rigid and winding filaments forming network, very distinctly articulated but without contractions, straight interstices hyaline, a single zoospore in each individual cell. It differs greatly from related ones in the absence of contractions and the presence of a single zoospore and in a way forms a transition to the Oscillatorineae.
Agonium centrale. Transparent filaments 3–5‴ long [= 0.03–0.05 inch, or 0.76–1.27 mm], starting from a common center, attached to stones, with cells twice as long as they are wide, oval muddy / cloudy / dark zoospores.
2. *Observat. au Microsc.* Pl. 13. fig. 14.
3. General characteristics. Flat rounded cells completely without sides and boundaries, zoospores 5–6 radiating from a common midpoint. *Homalodiscus vulgaris*, with round cells, zoospores 5–6. *Homalodiscus ovalis*, oval cells indistinct zoospores.
4. Dujardin, *Observat. au Microsc.* Pl. 30 fig. 20.
5. *Act. Acad. Berol.* (1841), p. 378.

– *Tessella arcuata* Her. – *Tessella interrupta* Ehrb. – *Tessella Ralesii* nob. (*Tessella catena* Rales, Jardines. *Mag. Of nat. Hist.* Aug. (1840) Pl. 2 f. 1. – *Synedra ulna* Ehrb. (*Frustulia obtusa* Lgb.) – *Synedra capitata* Ehrb. – *Navicula gaillonii* (*Echinella fasciculata* Lgb.) – *Licmophora gracilis* – *Licmophora abbreviata* Ehrb. – *Podosphenia cuneata* Ehrb. – *Achnanthes longipes* Ag. In the swamps at Toldboden close to Copenhagen. – *Achnanthes brevipes* Ag. – *Achnanthes intermedia* – *Achnanthes minutissima* Ehrb. (*Bangia rutilans* Lgb., *Nannema Hofmanni* Ehrb.) near Copenhagen (Toldboden in the swamp areas, Ny Badehus) – *Acineta tuberosa* Ehrb. (Ny Badehus).

The area with brown algae.

(On the plate it is marked with both light and dark brown color).

§18. We have already seen above that the area with brown algae is located outside and deeper than the area with green algae at a depth from 3–5 fathoms to a depth of 7–8 fathoms. It is easy for anyone looking at Plate I, where this area is marked in brown, to see that it is the most extensive part of the Sound, and that it is not lacking anywhere; the depth of the sea is suitable for these plants, and the bottom consists of sand mixed with larger and smaller stones. This area can be divided into two subareas, a) a subdivision with fucoids and *Zostera marina*, b) a subdivision with Laminaricae.

The sub-area with fucoids and Zostera marina.

(On the plate marked in light brown).

This sub-area is the part of the area with brown algae, which is closest to the shore. It occupies most of the bottom, and in some places extends several miles from the shore, such as off Skovshoved. They are like the savannas of the sea, for *Zostera marina*, which is dominant here, resembles grass so much that it is called seagrass by the fishermen, and over large areas gives the bottom the same uniformity that is peculiar to the tropical savannas. This uniformity is interrupted either by bare areas where the sand lacks any kind of vegetation, or by scrub growth with *Fucus vesiculosus* or *Fucus serratus*. The diverse nature of the bottom results in either species of the genus *Fucus* or *Zostera marina* dominating; where sand is mixed with few rocks, these fucoids dominate, as well as in the middle part of the Sound, where there is a large amount of rocks, as well as also in the northern and southern part. In addition to the species of *Fucus* mentioned earlier, these brown algae are specific to this area: *Halidrys siliquosa, Desmarestia aculeata, Desmarestia viridis, Stilophora tenella, Chorda filum, Chordaria flagelliformis, Stilophora nodulosa, Leathesia marina.*

NOTE 1. The fortress Trekroner protrudes in this area like a rocky rampart, as it is built as an artificial rock of stone lowered to a fairly great depth, where algae, which are brought there by various currents, find the place that is most appropriate to their natural requirements. Therefore, one can find here in the different seasons a large number of new and rare species, especially of brown algae. Here, at any time of the year, a diversity of algae can be observed in a very good way, as different species appear almost every month. Others can be found all year round, such as *Cruoria pellita*, *Zonarina liebmanni*, and others. Of other rare species that occur here, one should note: *Lichina confinis*, a thin variety of *Petalonia fascia*, *Scytosiphon lomentaria*, *Ectocarpus siliculosus*, *Elachista fucicola*, *Stilophora nodulosa*, *Mesogloia vermiculata*, *Nemalion multifidum*, *Dumontia contorta*, *Urospora penicilliformis*, *Rivularia nitida*, *Lyngbyea lutescens*, *Gallionella moniliformis*, Agon. Of lichens, *Verrucaria maura* occurs.

NOTE 2. Of algae that are predominant in the following area, the following, which do not occur infrequently in this area, should be noted: *Polyides rotundus*, *Furcellaria lumbricalis*, *Rhodomela confervoides*, *Ahnfeltia plicata*, and species of *Chondrus*.

The sub-area with kelp, Laminaria.

(On the plate marked with dark brown color).

This sub-area, the outer part of the brown algal zone, continues for the most part into the following. It is not as extensive as the previous sub-area, as it is completely missing in quite a few places. Its extent can easily be seen on plate I, where it is marked in dark brown. It corresponds to the part of the Sound which, due to the natural constitution of the bottom, has been named the stone area. The extent is greatest on the Scanian coast, where it stretches almost without interruptions from Kullen to Landskrona; on the Danish coast it has clear boundaries only to the south from Helsingør. This sub-area can be considered the forest of the sea. Laminarians, 10–15 feet tall, are pressed together upright like the trees of the forest, viz. *Laminaria digitata* [including its synonym *L. latifolia*], *Saccharina latissima.* Places with laminarians particularly occur to the south off Barsebäck; from this it appears that they are closer to the southern part of the Sound than previously stated.[1]

Listing of brown algae.

§19. Fucoids.

Halidrys siliquosa Lgb. – *Fucus vesiculosus* L. – *Fucus serratus* L. – A thin variety of *Ascophyllum nodosum* at Kullaberg.

Lichens.

Lichina confinis Ag Trekroner. Helsingør Hornem.

Kelp.

Laminaria digitata [including its synonym *L. latifolia*] – *Saccharina latissima* – *Petalonia fascia* Fr. in fl. Scan [Fries, E. 1835. Corpus florarum provincialium Sueciae. I. Floram Scanicam] (*Laminaria fascia* Ag.) Trekroner.[2]

Sporochnales.

Desmarestia aculeata Lamrx. Kullaberg. – Tårbæk. – *Desmarestia aculeata* var. *complanata* Ag. Together with the previous. – *Desmarestia aculeata* var. *plumosa.*

---

1. I. Agardh, *Novitae florae Sueciae ex Algarum familia*, p. 3.
2. *Laminaria Fascia* var. *tenuior* does not differ, as I. Agardh thinks (*Novitae florae Sueciae ex Algarum familia*. p. 15).

Hornem. (*Ectocarpus densus* Lgb.) between Helsingborg and Landskrona *Desmarestia viridis* Grev. Kullen, Hven.

Chordaceae.

Chorda filum Lamrx., *Scytosiphon lomentaria* Lgb. Trekroner but first 1842. *Asperococcus fistulosus* Grev. – *Punctaria tenuissima* (not *Laminaria Fascia* var. *tenuior* Lgb.) in between seagrass (*Zostera marina*). – *Dictyosiphon foeniculaceus* Ag. Kullen. – Trekroner. Is it a separate species? However, I have not been able to distinguish it from some different forms of *Chordaria flagelliformis*. – *Ralfsia fungiformis*, in swampy areas and between stones at the fortress Trekroner.[1]

1. This plant has long been known by all algae researchers, but there are striking mistakes when it comes to interpreting it. It has been discovered by Lyngbye by the rocks in the Norwegian Sea, and he has attributed it to *Zonaria deusta Flor. Dan.* (= *Ralfsia fungiformis*). It has thus been properly determined; however, the famous Agardh is in doubt and reckons it for a *Crusta bacilaris* – a kind of lichen. When I first saw this plant, I thought it was related to the genus *Zonaria* and did not easily understand how Lyngbye could have attributed it to *Zonaria deusta Flor. Dan.*; that this is the case, however, I have been persuaded by the authority of Hofman Bang, who in this case is of more or less the same opinion as Lyngby. And Liebmann, who has returned from a journey to Mexico, has confirmed this view with authentic samples. The very diverse family, related to *Zonaria*, is easily distinguished at first glance by the fact that the upper part grows very close to rocks in the tidal areas. Many other traits show up for the one who examines it further. Since there are samples of this plant in many herbariums, they can therefore be easily examined, I will here simply refer to the description and depictions that Liebmann will make sure to be made in *Flor. Dan.* according to the samples that I have collected. –

Ectocarpaceae.

Cladostephus spongiosus f. verticillatus Lg. close to Helsingør Liebmann. – *Ectocarpus siliculosus* Lgb. Trekroner. – *Pylaiella littoralis* Lgb. [including its synonyms *Ectocarpus compactus* and *E. brachiatus*] – *Spongonema tomentosum* at the island Hven. – *Sphaceloderma caespitulum* Trekroner. – *Sphacelaria cirrhosa* Ag. at Kullaberg. – *Chaetopteris plumosa* Kullen Hven. – *Elachista fucicola* Fr. – *Elachista stellaris* Areschoug in *Linnea*, vol. 16, p. 233 at Kullaberg? *Elachista fucicola* at Trekroner[1]). – *Cruoria pellita* Fr. in fl. Scan., *Erythroclathrus pellitus* Liebm. in Krøyers *Tidsskrift*, vol. 2, p. 189.

Chordarieae.

Chordaria flagelliformis Ag. Kullen-Trekroner. – *Stilophora nodulosa* Trekroner.[2] *Stilophora tenella* (*Chordaria rhizodes* and *C. paradoxa* Lgb.) – *Leathesia marina* (*Chorynephora marina* Ag.)

---

1. This species is easily distinguished from the others in this genus both by the hemispherical dome-shaped form that can be found and by the fact that most of the filaments are attached together to a dense gelatinous mass, and only a very short part of them is thus free. Sometimes, however, one encounters very long free threads, and it thus comes to resemble *Elachista fucicola*. See *Flor. Dan.*, an unpublished plate.

2. This rare plant occurs almost all year round at Trekroner. Lyngbye's species name should be preferred to Agardh's, because it is first described in *Flor. Dan.* under the name *Ceramius tuberculosus* (= *Stilophora nodulosa*), later it is called *Chaetophora nodulosa* (= *Stilophora nodulosa*) by the famous Agardh. Perhaps this species and *C. rhizodes* should be attributed to their own genus.

The area with red algae.

(On the plate marked in purple).

§ 20. This area is the outer limit of the vegetation of the Sound. Red algae are characteristic of a depth of 8–20 fathoms. This area of the Sound is only clearly delimited in a few places and merges here with the outermost sub-area of the previous area with kelp. A rocky zone is favorable for red algae as well as for kelp, but where the external conditions are appropriate, they are distributed to a greater depth. No place of the Sound is so deep that red algae are unable to grow there, as long as they have a bottom suitable for them with either stones or shells; but as the greater part of the Sound outside the rocky zone is covered with very soft clay, they seldom appear here. The famous Agardh has rightly already stated that the places where red algae and green algae grow are opposite to each other, as the former particularly dominate in the northern part of the Sound, the latter in the southern. But red algae occur closer to the southern part of the Sound than this very prominent algae researcher thinks; many red algae are found to the north of Hven, and also at Stevns three species of *Delesseria* and others appear to a large extent together. Species particularly characteristic of this area are: *Dilsea carnosa*, the species of *Delesseria*, *Polysiphonia* [*Hutchinsia*], *Callithamnion*, *Ceramium*, *Gigartina* and *Odonthalia dentata*.

NOTE. From what has been said, it is easy to understand that in the various areas there are not only the algae of the type from which the name is given; in all areas, algae representing all sections occur, but in each area each section contains so many of the characteristic species that the others almost seem to disappear because of their dominance. The names are thus derived from plants that are particularly prominent for each region.

Enumeration of red algae.

§ 21. Chordariaceae and Nemaliaceae.

Mesogloia vermicularis Ag. Kullen-Trekroner. *Eudesme virescens*. (*Mesogloia Zosterae* Areschoug *Algae Scandinaviae Exsiccatae*, part III, 67) Helsingør Liebm. Copenhagen. – *Helminthora multifida* Fr. (*Chordaria multifida* Lgb.) Kullen-Trekroner.

Dumontiaceae.

Dumontia contorta (*Gastridium filiforme* Lgb.) Hven-Trekroner. – *Dilsea carnosa* (*Halymenia edulis* Ag.) Kullen-Helsingb.

Polyidaceae.

Polyides rotunda Grev. Helsing.

Furcellariaceae.

Furcellaria lumbricalis.

Florideophyceae.

Delesseria sanguinea Lmrx. – *Phycodrys rubens* – *Delesseria alata* Lmrx. – *Membranoptera alata* – *Odonthalia dentata* Lgb. Kullen-Hellebæk. – *Rhodomela confervoides* – *Cystoclonium purpureum* Kullen-Hven. – *Gracilariopsis longissima* Together with the previous. – *Ahnfeltia plicata* – *Dictyosiphon foeniculaceus* (*Scytosiphon hippuroides* Lgb.) at Hellebæk. – *Chondrus crispus* Lgb. [including its synonym *Chondrus crispus var. incurvatus*] at Hellebæk. – *Phyllophora pseudoceranoïdes* – *Coccotylus truncatus* [including its synonyms *Chondrus Brodiaei angustissimus* Suhr and *Chondrus Brondiaei ligulatus* Suhr] at Hellebæk. –*Gloiocladia repens* at Hellebæk. – *Ptilota gunneri* at Helsingør Liebm. At Kullaberg.

Ceramiaceae.

Leptosiphonia brodiei at Helsingør Liebm. – *Leptosiphonia brodiei* The same location. – *Vertebrata byssoides* at Kullaberg. – *Leptosiphonia fibrillosa* [as *Hutchinsia tenuis*] at Lomma Agardh. – *Vertebrata fucoides* [including its synonym *Hutchinsia violacea* Lgb.] – *Polysiphonia stricta* (*Hutchinsia stricta* Ag. in Syn.) Hellebæk, Kullaberg. – *Leptosiphonia fibrillosa* [as *Hutchinsia divaricata*] at Helsingør Liebm. – *Polysiphonia urceolata f. lepadicola* at Helsingør Liebm. – *Ceramium virgatum* – *Ceramium diaphanum* Roth. – *Ceramium ciliatum* Ducluz. – *Carradoriella elongata* Kullen-Landskrona. – [including its synonyms *Ceramium elongatum var denudatum* Ag. (*Ceramium brachygonium* Lgb.) and *Ceramium elongatum proliferum* Ag.] Together with the previous. – *Gaillona hookeri* Helsingør. Lieb. – *Rhodochorton purpureum* at Kullaberg I. Agardh. – *Callithamnion tetragonum* at Kullaberg. – *Colaconema daviesii* on *Ceramium virgatum* – *Spermothamnion repens* Fries in fl. scan. – *Compsothamnion thuioides* at Kullaberg Ag. Fr.

§ 22.[1] When we consider the horizontal distribution of the algae, we easily see that it extends only along the longitudinal direction of the Sound, viz. from north to south, but not from east to west, as it is the same as the vertical distribution. This is because the largest part of the Sound is formed by two smooth slopes, which form a gorge between Zealand and Scania, and the part that lies in between lacks, as we have already shown, any kind of vegetation. The sound provides a very good opportunity to observe how the vegetation in the direction from south to north is affected in different ways by the salty sea and by the less salty; as we have seen before, the Sound is one of the places where the salt water from the Kattegat and the almost fresh water from the Baltic Sea mix. Therefore, red algae dominate in the northern part of the strait, brown algae in the middle, green algae in the southern. The diversity of the distribution is indicated on plate 1 with the corresponding colors.[2] However, the red algae, as if compressed, relinquish their dominance of the sea. They tend to shrink into dwarf forms rather than disappear entirely. The forms

1. [Called § 21 because of a printing error.]
2. Viz., purple red, olive and green.

of the genera *Delesseria*, *Chondrus* and others, which also occur at Stevns, are small and almost green, so that one may at first doubt whether they belong to the same kind.

On the conditions that form the basis for the distribution of algae by regions.

§ 23. The way in which algae have generally been collected has not allowed a demonstration of the general laws of geographical distribution, as all species are collected after being brought together on the shore and therefore have been torn away from their original places. J. Agardh was the first to give an excellent example in a book that I have often cited. He states the following conditions, which form the basis of the distribution of algae: 1) the chemical nature of the sea, 2) the depth of the sea, 3) the changing stillness and turbulence of the sea, 4) the nature of the substratum. Since algae come into contact with two midpoints, the bottom and the surrounding sea, the conditions that affect them can be divided according to:

1. Bottom conditions
   a) the chemical nature.
   b) the mechanical nature.
2. Surrounding sea conditions
   a) the chemical nature.
   b) the depth
      aa) the nature of the light.
      bb) the nature of the turbulence.

§24. The fact that the chemical nature of the soil cannot be a factor that forms the basis of the algal vegetation is easily recognized, as the algae's roots simply have the function of attachment and not receiving nourishing liquids. Thus, the chemical composition of the bottom, being dissolved by the surrounding seawater, can thus only influence the system indirectly.

The physical conditions of the bottom, on the other hand, are of great importance in forming the basis for the places where algae can grow according to the diversity of the areas where they are grouped. Algae can only attach roots on firm substrate; hence this law states that algae are lacking in all places in the Sound where the bottom lacks stones. But since a smaller part of the Sound consists of rocks or gatherings of stones, and a larger part consists of sand or shells, it is therefore clear that a smaller part of the bottom is covered with algae, and that it is likewise the special place they grow, where there is a very large quantity of stones; we have already seen that it is also in a zone of rock where both red and brown algae have developed the greatest amount of vegetation. *Zostera marina*, on the other hand, which does not belong to the algae, can only attach roots in loose sand and has established its vegetation in sandy zones. As *Zostera marina* occupies the largest part of the Sound, it surpasses all algae in biomass. Neither *Zostera marina* nor algae can develop roots in the very soft clay, and therefore there is a zone consisting entirely of clay without any vegetation. The chemical conditions of the sea are of great importance in determining the appearance of its vegetation; it is sufficiently expressed[1] in the difference between the vegetation in the northern and the southern parts of the Sound. The varying salinity in the different parts of the Sound means that green algae dominate over the red algae, which almost disappear in the southern part.

After the chemical nature of the water, there is hardly anything that is as important to the algae as the difference in light caused by different depths. The power of this light has not yet been sufficiently studied, as it has only been said that the intensity of light changes according to the difference in depth;[2] but we will now see that other factors can also be important. Optical science shows that rays of different colors, of which a light without color is composed, which penetrate a liquid that is not completely transparent (which is the nature of the sea water) are refracted in different ways, so that not all penetrate equally deep through the ocean. Red rays reach the greatest depth, then the orange ones, finally the others in a regular order, so that the blue and purple ones penetrate the least. We know this property of the light from the sunlight in the morning and in the evening, when the red and yellow rays of the sun penetrate the air, which is full of heat. From this it is understood, as the experience of divers also shows, that almost exclusively red rays reach the greatest depths of the sea. But in the shallower areas we see that areas with algae,

1. See J. Agardh. *Novitae florae Sueciae ex Algarum familia*, p. 4.
2. See J. Agardh. *Novitae florae Sueciae ex Algarum familia*.

which are of different colors, grow at different depths; when we enumerate algae in accordance with the order in which they occur from the surface of the sea to a greater depth, and also measure the rays according to the order in which they penetrate deeper, we therefore see that we find areas of algae that are of different color in the places where we above have shown that there is the greatest amount of rays of the corresponding color.

Color	Algae	Depth
Purple rays Dark blue rays Light blue rays	Blue-green algae [Cyanobacteria]	At the surface.
Green rays	Green algae [Chlorophyceae]	10–25 feet.
Yellow rays Orange rays	Brown algae [Phaeophyceae]	25–50 feet.
Red rays	Red algae [Rhodophyceae]	50–65 feet.

We by no means claim that we will have solved the very difficult phenomena that the different colors of the algae exhibit; for we should then explain why algae have the same color as the rays of the ambient light; but we believe that with the preceding remarks we have moved closer to an explanation.

In addition to the importance that depth has for determining the nature of light, it also has a very large impact on determining the influence of the waves on the bottom, as their influence is inverse to the depth. From this appears the general law that green algae grow where the violence of the waves is greatest, as we meet them at the shallowest depth, then brown algae and finally the red algae, where the violence is very small. But besides the motion of the sea, which is caused by the waves that do not touch a greater depth, there is another, which is caused by the current, which is often very great, where the motion of the waves is very small. This is the case in the Sound, where kelp and red algae are found. This depth is so great that one cannot observe movement of the waves on the bottom; but the force of the current is also very great here. From this it is understood that algae for which the current is favorable can disappear in the sea which is exposed to the violence of the waves, and vice versa. It should be noted that the movement of the current is also indirectly beneficial to the algal growth, as it makes the nature of the bottom very suitable for it; therefore, it is difficult to determine whether the nature of the bottom or the current is most beneficial to algal growth. This table provides an overview of the conditions that form the basis of the algal vegetation.

Salinity	large	Red algae
	less	Brown algae
	little	Green algae
By depth: Light intensity	large	Green algae
	less	Brown algae
	little	Red algae
By depth: Violence of the waves	large	Green algae
	less	Brown algae
	little	Red algae

§ 25. Agardh was the first to show that very important characteristics for the algae, which are the basis for their subdivision, can be deduced from their colors. This basis for the division is today accepted by everybody, so that there is hardly anyone who doubts that the division of algae into green, brown and red is in accordance with nature to the highest degree. However, it has never been explained why algae in this respect differ so much from other plants, why color, which is otherwise the characteristic of the species, is a trait of a very significant difference. Above we have seen that algae that have different colors are not affected by light in the same way when they grow at different water depths or in different areas; this very important property can only be understood by this relationship. The algae are the only order whose different suborders are differently affected by the light, so one should not be surprised that they differ from the other plants in the property of the color. The reason why the color is very important when they are to be systematically subdivided is thus found in the special habitats of these plants. That this subdivision is consistent with nature is also clear from the fact that these three suborders are composed of analogous parts; here, as always in every natural system, we find an evolution in parallel series with analogous parts; this basis for a subdivision has hardly ever been used in a system with as much consistency as by Agardh in the distribution of algae.[1] We show an overview of the algal families based on these principles:

Order Algae.

1. *Suborder* Phaeophyceae.
   1. *Family* Fucaceae.[2]
   2. *Family* Laminariaceae.
   3. *Family* Ectocarpaceae.

2. *Suborder* Rhodophyceae.
   1. *Family* Dumontiaceae [called Florideae].
   2. *Family* Halymeniaceae.
   3. *Family* Ceramiaceae.

3. *Suborder* Chlorophyceae.
   1. *Family* Ulvaceae.
   2. *Family* Cladophoraceae [as Confervoideae].
   3. *Family* Diatomaceae.

---

1. *Species Algarum*, p. LXXII.
2. Including Sporochonoideae, Dictyoteae and Chordaricae.

In the first two suborders (brown and red algae), the first two families (Fucaceae and Dumontiaceae) show similar distribution in all dimensions and resemble each other in their substance; the other two (Laminariaceae and Halymeniaceae) show a distinct distribution in two dimensions and resemble each other in their surfaces, the third (Ectocarpaceae and Ceramicaceae) show a distinct distribution in one dimension and resemble each other in outline. Thus, in the lowest suborder, there is no member sufficiently developed to correspond to fully formed bodies; instead, the organisms present are limited to sheet-like and filamentous forms (e.g. Ulvaceae and Confervoideae). Conversely, this suborder also contains an entirely new group of such simplicity that it lacks any analogue in higher taxa, consisting merely of point-like entities (the Diatomaceae). Therefore, I believe that Diatomaceae should be classified as a kind of green algae, although most correspond in color to the brown algae; there is no doubt that they represent the lowest structure among the algae.

NOTE. The number of all plant species occurring in the Sound is one hundred and seventy-three. In addition to one hundred and sixty-six species of the algae order, these plants are found here: by Naiadeae: *Ruppia maritima* [including its synonym *Ruppia rostellata*], *Zannichellia palustris* L., *Zostera marina* L., by Characeae: *Chara Baltica* Fr. and Aspg., *Chara nidifica*) *Engl. Bot.*, by Lichineae: *Verrucaria maura* Fr. The species are distributed as follows between the different divisions of algae:

Brown algae	36
Red algae	41
Green algae	51
Diatomaceae	38
[Total]	166

CHAPTER 4.

About the areas with animals in the Sound.

§ 26. In the foregoing remarks, I have already drawn attention to what is to be understood by an area, and I will therefore immediately begin to examine what has been written so far in this subject concerning the geography of marine animals. Audouin and Edwards, who have studied the Franco-Gallic coasts, were the first to observe the great diversity of the animal kingdom at different depths in areas exposed by the tide; based on this diversity, they have divided them into four areas.[1] The same conditions have been found by Sars on the coasts of Bergen.[2] The areas are as follows:

1. *The area with barnacles* covers the upper part closest to land. In addition to barnacles, *Nucella lapillus* in particular dominates here.
2. *The area with the limpets*, which is the lower part; of this the limpets [*Patella pellucida, Testudinalia testudinalis, Tectura virginea*], *Turbines* [= *Rissoa*], *Mytilus*

1. *Annales d. scienc. Natur.*, vol. 21, p. 26. [should be page 326].
2. *Beskrivelser og Iagttagelser over nogle mærkelige eller nye i Havet ved den bergenske Kyst levende Dyr* (1835), p. VI.

edulis, sea anemones [Actiniae] are characteristic; of plants it is seaweed [*Fucus* spp.].

3. *The area with Corallinae.* Here are found primarily *Corallina officinalis*, *Modiolus modiolus*, ascidians, sponges, bryozoans, annelids, and of plants *Zostera marina*.
4. *The area with kelp.* Nudibranchs, sea stars, skeleton shrimps, sea spiders dominate. None of these authors, however, have examined the distribution of animals at a greater depth. However, when we compare the nature of the coasts of Franco-Gallia and Norway that have been studied so far, with those which we will now subject to study, we would immediately expect a very large diversity in the distribution of animals due to their very diverse nature. These shores are steep and rocky, of which large areas are either covered by the tide or are exposed; the very shallow areas are not exposed to the influence of the sea, which alternately grows and becomes smaller [tidal movements]. It is therefore easy to understand that with a gentler inclination of the shores, the longer each area is; the same difference in the depth of the sea as that which appears on a steep coast over a much smaller area, appears here just over a larger area. As we will see below, if we imagine the four areas that we have previously mentioned extending over a larger area, they largely correspond to those found in the Sound.

§ 27. I have previously pointed out that the distribution of animals in the areas of the Sound cannot be explored by direct observation, such as at steep shores, a large part of which are exposed at low tide, but almost everywhere the animals that live on the bottom can be studied by the use of a special instrument. In this way, by examining the bottom of the Sound across its entire length from the Danish to the Scanian coast, I have learned that at different depths there are different species, and even that there are entire genera and families that are special for the different depths. Based on these differences, the bottom can be divided into different areas; their names can be deduced from the animals that are characteristic for them, particularly from those whose general divisions (i.e., orders and families) correspond to the diversity of places; that is, of the animals, not only different species and genera, but also families or orders that appear with considerable difference according to the external circumstances. As for the animals in the Sound, I think it is accurate, at least as far as Mollusca is concerned. Of the three areas that will be explained below, *the area with Trochus* [trochoid snails], adjacent to the coast, occupies the whole zone with sand; outside this zone, *the area with nudibranchs* within a rocky zone and the area of kelp [*Laminaria*] and red algae collide; finally, *the area of whelks*, which covers a zone of shells, can also be called the deep area.

The area with Trochus [trochoid snails].

(On the plate marked in blue, green, light brown and white).

§ 28. This area extends from the shore and as far as the sandy bottom exists, i.e. from 0 to 7–8 fathoms. Kullen is the only place where it is almost completely missing because the depth is great close to the shore. The general nature of this area is already largely apparent from the foregoing. From this it is evident that the force of the waves here is very violent. The movements of the waves can reach the bottom everywhere in this zone and thus also the animals that live on it; but the effect on the bottom corresponds inversely to the depth. It is therefore necessary that all animals that live on it can somehow defend themselves against the dangerous power of the waves. We will see that this can happen in two ways. In the animals, which are surrounded by a calcareous shell, it evokes thickness and great hardness; a proof of this is first and foremost *Littorina littorea*, which is often carried by the waves between the rocks of the coast without any damage. Animals that lack shells, nature has made sure are defended in other ways; they can hide in burrows dug in the sand, so that they not only avoid the movement of the waves, but also the force of the air, which will affect them when the waves retreat. Something special for this area is that a large part of it is exposed to the air when the sea retreats, even though it is only in the northern part that tidal movements occur; however, the depth of the sea changes according to different currents. Among the animals that hide in this way are: *Mya arenaria*, *Hediste diversicolor*, *Arenicola marina* and *Corophium volutator*.[1] A rich vegetation, which is characteristic of this zone, and its great diversity, is of importance to the animals. Like the green algae *Ulva* spp., which grow with its associated forms [animal(s)] (such as *Peringia ulvae*, *Procerodes littoralis*), so the dense assemblage of fucoid brown algae has its own associated forms (such as *Platynereis dumerilii*) and grass meadows that, with a wide distribution, are beneficial to others than *Zostera marina* (i.e., *Nicolea zostericola*, *Platynereis dumerilii*).

The animals of this area can be described in these few words: *most animals here are either phytophagous or possess adaptations for protection, such as a hard shell or the ability to bury themselves in the sand to withstand the destructive forces of the waves and exposure to the air.*

The particular species found almost everywhere in this area should be mentioned: *Crangon crangon* and *Praunus flexuosus*, which occur mainly along the coast. *Palaemon adspersus* occurs in large numbers together with *Idotea balthica* at a slightly greater depth; in quieter places *Jaera albifrons* and *Corophium volutator* dominate. *Talitrus saltator* and *Orchestia littorea* live hidden between dense populations of brown algae (*Fucus* spp.). *Semibalanus balanoides* covers rocks on the shores of the sea. *Hediste*

---

1. See: Krøyers *Naturhist. Tidskr.*, vol. 3, p. 558; *Beretning om en Excursion etc.* by Ørsted.

diversicolor lives in burrows under rocks, always in very large densities. Sand tubes of *Spio seticornis* appear vertically in large quantity on even shores. The excrements of lugworm (*Arenicola marina*) twisted in spirals at their openings, are a sign of large numbers in their burrows. *Lumbricillus lineatus* and *Lineus longissimus*, which form clumps, occur under any of the coastal rocks. *Littorina littorea*, *L. obtusata* and *L. fabalis* prefer to live on rocks, especially when they are not covered by water, rather than submerged. *Peringia ulvae*, *Mytilus edulis*, *Cerastoderma edule*, *Limecola balthica* and *Mya arenaria*, occur at slightly greater depth. *Echinus esculentus* is found exclusively in the outer part of this area, i.e., at a depth of 5–8 fathoms, together with *Asterias rubens*. *Obelia geniculata* and *Membranipora membranacea* which cover plants and pilings.

NOTE 1. This area can be divided into several subregions; however, since these are very closely connected with one another, they possess not entire families as in the regions, but only genera or species of their own. Thus, although *Littorina* spp. often occur in the outermost area, they stay especially in the area closest to the shore, so they can almost be considered amphibious, as they live as much in the water as above it. Therefore, one might be able to set up following sub-areas.

1. A sub-area with *Littorina* spp. covers an area of the shore which emerges when the water recedes. In addition to *Littorina* spp., the species *Nereis*, *Spio*, *Arenicola* and *Mya*, mentioned above, are specific to this sub-area.

2. A sub-area with *Mytilus edulis* extends from the outer boundary of the previous one to the areas where sand is mixed with shells. Although *Mytilus edulis* is also found in the previous sub-area, it is dominant here. Particular species are: *Akera bullata, Ciona intestinalis, Carcinus maenas, Nicolea zostericola, Echinus esculentus* and *Lucernaria quadricornis*. Among plants, seagrass, *Zostera marina*, dominates here.
3. A sub-area with *Tritia reticulata* covers the outer part of this area, where sand is mixed with shells and devoid of vegetation. This sub-area is found only where a rocky area is missing, which, if it occurs, forms a clear boundary between a sandy zone and one with shells. If it is missing, as in almost the entire middle of the Sound, this area's species and those of the depths are mixed together. I can only mention two species as specific to this area: *Tritia reticulata* and *Corbula gibba*.

NOTE 2. In this area there is a special bottom with characteristic animal species in some places: mud, which is the same in the sea as soil on land; it arises from the decay of plants that have been dissolved. It forms the basis of a very soft and very rich black mass, which is particularly favorable for *Ulva lactuca* and *Ulva lactuca* var. *latissima*. It is formed in deeper places where accumulated plants decay and is not renewed by the water current. Species that are typical should be mentioned: *Philine* (*aperta*?), *Amphictene auricoma, Lumbricillus lineatus, Lineus ruber* and *Tetrastemma melanocephalum*.

NOTE 3. Although fish among all sea creatures are least restricted to certain places as they swim around freely, some are specific to this area. Of these must first be mentioned: *Gasterosteus aculeatus*, a small but strong and powerful fish that swims close to shore and initiates strange battles; *Spinachia spinachia*, which follows shoals of *Palaemon adspersus*, and eats the eggs they produce; *Myoxocephalus scorpius*, is dull, confused, and ferocious, and moves slowly toward the base between the leaves of the *Zostera marina*; *Pholis gunnellus*, which lives quietly between rock and brown algae (*Fucus*) and sometimes, when the water recedes, is left on the dry land; *Zoarces viviparus*, which also resides between stones and brown algae (*Fucus*), and while the stomach grows because of a large number of young, it is nourished by blue mussels (*Mytilus edulis*); in addition are there European flounder (*Platichthys flesus*), which sometimes enters the streams, lesser sand eel (*Ammodytes tobianus*), and eel (*Anguilla anguilla*).[1]

NOTE 4. When comparing this area with the Franco-Gallic and Norwegian coasts, it is easy to understand that it corresponds to the former areas of barnacles (Balanidae), limpets (*Patella, Testudinalia, Tectura*) and calcareous red seaweed (*Corallina*). However, this difference is particularly clear as limpets (*Patella, Testudinalia, Tectura*) and calcareous red seaweed (*Corallina*) must be considered special in the Sound in the following sub-area.

List of all animal species in the area of Trochus [trochoid snails].

§ 29. **Crustacea.**
Geryon trispinosus Gilleleje, Lyngbye. – *Carcinus maenas* Leach. – *Pinnotheres pisum* Latr. Gilleleje Lyngbye. – *Lithodes maja*. In the northern part of the Sound Krøyer. – *Crangon crangon*. – *Hippolyte gaimardii* Edw. Hellebæk. – *Palaemon adspersus*. – *Praunus flexuosus* Lamk. – *Talitrus saltator* Edw. – *Orchestia littorea* Leach. – *Hyperoche medusarum* Kr. in *Aurelia aurita*. – *Hyperia new species*? Together with the previous in *Aurelia aurita*

1. See Krøyer *Danmarks Fiske*, parts 1–3.

– *Gammarus sp.* Hellebæk-Hven. – *Gammarus locusta* Fabr. – *Idotea pelagica* Leach. – *Idotea balthica.* – *Idotea emarginata* Fabr.[1]) – *Cyathura carinata* Kalvebod beach. – *Heterotanais* oerstedii Kr. [and its synonym *Tanais Curculio*]. The same locality. – *Jaera albifrons* Leach. – *Sphaeroma* new species?

Arachnida.

Trachheariae. Acaridae.

Thalassarachna basteri[2]) at Kalvebod beach [including its synonym *Acarus setosus* nob.[3])] at Trekroner.

Annulata.[4]

Harmothoe impar Ørsd. Tårbæk. – *Harmothoe imbricata.* – *Pholoe baltica* Ørsd. – *Platynereis dumerilii* Hellebæk [including its synonyms *Nereilepas variabilis* and *Nereis zostericola*] Hellebæk. – *Hediste diversicolor* Mü. – *Phyllodoce assimilis* Ørsd. Kullen. – *Phyllodoce mucosa* Ørsd. – *Polydora ciliata* between Copenhagen and Trekroner. – *Spio seticornis* O. Fabr. at Ny Badehus. – *Spio filicornis* O.

---

1. I have already expressed my doubts about the diversity of these three species in the past (see Krøyers *Tidssk.*, vol. 3, p. 561).
2. *London Mag. of Nat. Hist.*, vol. 9, p. 353.
3. With ash-gray body, oblong-oval, and contracted towards the anterior and posterior part, the posterior part very short tapered, sensory threads hidden under a blunt snout. With flippers of equal length, with numerous long hairs on the posterior part of the body, on the anterior part of the body just two; length 1/300 inch [= approximately 0.085 mm]. Both species are very distinct.
4. Most species of the order Annulati listed here are described in my book: *Conspectus Annulatorum Danicorum* (1843).

Fabr. Helsingborg. – *Arenicola marina.* – *Spirorbis spirorbis* – *Fabricia stellaris* (*Tabularia Fabricia Faun. Groenl.* Fig. 12, *Nais equisetina* Düges *Ann. D. sc. Nat.*, vol. 8, Pl. I. f. 24). – *Nicolea zostericola* nob. Issefjord. – *Amphictene auricoma* Sav. – *Tubifex serpentinus* nob. Close to Tårbæk[1]). *Lumbricillus lineatus*[2] [including its synonym *Lumbricillus verrucosus*] Kalvebod beach. – *Paranais litoralis* (Krøyers *Tidsskrift* vol. 3, p. 136.) Kalvebod beach. – *Nais elinguis* Müll. together with the previous.

Apoda.[3]

Dendrocoelum lacteum Ørsd. Copenhagen-Stevns. – *Planaria torva* Mül. together with the previous. – *Procerodes littoralis* – *Foviella affinis* Ørsd. Copenhagen. – *Archilopsis unipunctata* Ørsd. – *Monocelis lineata* Ørsd. – [including its synonym *Monocelis rutilans* Ehrb.] Kalvebod beach. – *Vortex mytili* Ørsd. – *Graffiellus croceus* [including its synonym *Prostoma suboviforme*] Kalvebod beach. – *Vortex littoralis* Ørsd. the same place. – *Dinophilus vorticoides* Tårbæk. – *Typhloplana marina* Ørsd. Hven. – *Convoluta*

1. The genus *Tubifex* differs from *Lumbricillus* particularly in the setae on the back, which are capillary bristles, and having barbs.
2. *Lumbricillus* [*lineatus*], the type of the genus *Lumbricillus*, is recognized by its dorsal and ventral bristles that are needle-shaped, shortened and approximately straight (see Kröyers *Tidsskr.*, vol. 3, pp. 130–31).
3. All species of the order Planariei listed in the following are described in my book: *Entwurf einer systematischen Eintheilung und specieller Beschreibung der Plattwürmer* (1844).

convoluta Kalvebod beach. – *Cephalothrix linearis* between Copenhagen and Trekroner. – *Tetrastemma subpellucidum* Ørsd. Snekkersten. – *Tetrastemma bioculatum* Ørsd. between Copenhagen and Trekroner. – *Tetra-stemma assimile* Ørsd. Kalvebod beach. – *Tetrastemma melanocephalum* Kalvebod beach. – *Cephalothrix rufifrons* between Copenhagen and Trekroner. – *Gibsonnemertes spectabilis* – *Lineus ruber* – *Nipponnemertes pulchra* between Copenhagen and Trekroner. – *Tubulanus annulatus* – *Pontonema muelleri*, barely *Enchelidium marinum* Ehrb. [Nematoda].[1]

Mollusca.

Gastropoda.

Lymnaea stagnalis Kalvebod beach. – *Peringia ulvae* [including its synonyms *Paludinella Baltica, Paludina Baltica* and *Paludinella vulgaris*][2] – *Eupaludestrina stag-*

1. *Act. Acad Berol.* (1835), p. 219. Since the *Vibrio marina* Mül. completely lacks eyes, it cannot be identical to the *Enchelidius marinus* Ehrb., as the famous Ehrenberg believes.
2. These two species, which often co-occur and are confused, are easily distinguished in this way:
Paludinella baltica, oblong-conical shell, almost perforated, impure greenish, dark, almost translucent, with curved convex whorls, forward spire not sharp, oblong foot, anteriorly truncated / hollowed, posteriorly rounded, but not very short. [Probably *Hydrobia ventrosa* [= *Eupaludestrina stagnalis*].]
Paludinella vulgaris, oblong-conical shell that is not perforated, translucent, shiny, dark-striped, whorls indistinct, less convex, forwardly sharp spire, oblong foot rounded anteriorly, posteriorly it becomes shorter. [*Peringia ulvae* (Pennant, 1777).]
At Kullaberg I have also found four species that seem to belong to this genus; however, I dare not decide it, as all the samples were without animals.

norum Kalvebod beach. Is it a new species of *Neritina fluvitalis?* – *Steromphala cineraria* Kullen-Landskrona. – *Littorina littorea* Fer. – *Melarhaphe neritoides* Kullen. – *Littorina rudis* the same locality. – *obtusata* – *Littorina fabalis* Turt. – *Lacuna vincta* – *Lacuna pallidula* Turt. Trekroner. – *Tritia reticulata* – *Nucella lapillus* Lamk. Kullen-Helsingborg. – *Bittium reticulatum* Issefjord. – *Akera bullata* Mül. – *Philine aperta* Trekroner. – *Doris lacinulata* Helsingør – Copenhagen. – *Limapontia capitata*[1] (*Planaria limacina* O. Fabr., *Planaria capitata* Müll.).

Acephala.

Mytilus edulis L. – *Cerastoderma edule* L. – *Corbula gibba* – *Spisula solida* L. – *Spisula* a new species? Kullen. – *Macomangulus tenuis* – *Limecola balthica* – *Mya arenaria* L.

Tunicata.

Ciona intestinalis Sav. – *Molgula manhattensis.*[2]

Echinodermata.

Echinus esculentus L. – *Asterias rubens* (*Asterias rubra* and *A. violacea* Mül.).

---

1. *Lond. Mag. of Nat. Hist.*, vol. 9, p. 79–80.
2. With spherical body, jelly-like, semitranslucent, greyish-white, with openings elongated into tubes, one of which is as long as the whole body, the other half or a third of this length; the opening of the latter is provided with notches of a length of 4 lines [= 4 hundredths of an inch ≈ 1.016 mm], and the former of a length of 5 lines [≈ 1.27 mm].

Polypi.

Lucernaria quadricornis L. – *Clava multicornis* On the beach itself. – *Alcyonium digitatum* – *Obelia geniculata* – *Membranipora membranacea* L.

The area with Nudibranchia.

(Marked with dark brown and purple on the Plate).

§ 30. This area has only a small extent in the Sound, yet a very large number of plants and animals occur here. Its boundaries correspond to the area that, due to the nature of the bottom with stones and shells and due to the vegetation, is called the area of kelp and red algae. It is thus completely lacking in the southern part of the Sound but dominates in the northern. It is thus provided with an extensive and species-rich vegetation, and as something special for this area, the water is renewed by a constant current. It thus differs from the other areas by the nature of the external forces that affect it. The bottom, which is rocky, prevents animals from penetrating it; however, this is not as necessary for them as in the previous area; here they do not have to fear the movements of the waves or the influence of the air. They can live safely between the large leaves of kelp. It is therefore no wonder that the animals differ very much from the animals of the previous region. I have previously presented the opinion that the vegetation of this area in a way resembles the forests on land. If you observe the animals as well, the tropical forests are also of interest. This is because there is not the same uniformity as on the vast meadows with *Zostera marina*, but a very large amount, richness and variety of animals and plants. Here, as in these forests, the bottom is so densely covered with plants that most of the animals in it must be mentioned. Skeleton shrimps, sea spiders, nudibranchiates, limpets, chitons (= Polyplacophora), sea squirts, sea cucumbers and sea anemones compete with the plants between which they move, in the splendor and variety of the colors. *Metridium senile*, *Stomphia coccinea*, *Urticina felina*, along with *Psolus phantapus* are reminiscent of the petals of garden flowers; *Aeolidia papillosa*, characterized by different colors, *Doris pseudoargus* with a color of rust-stained spots, *Polycera quadrilineata* with yellow lines, *Elysia viridis* characterized by its black color, and many other species that swim around between brown kelp fronds, and between blood-red *Delesseria*, *Ptilota*, *Odonthalia*, delight the eyes with a great variety in color and shape, and *Patella pellucida* shows an almost tropical color with the sheen of the rainbow. We can thus with these few words describe the general character of the animals of this area: most animals that occur here do not in color fall far behind the plants among which they reside, move slowly in between[1] or are almost attached to, and almost surpass them in variety; a constant current brings food, since they do not eat plants. They usually lack a hard shell,[2] but they are not exposed to the violence of the waves.

---

1. Also, Crustacea (such as *Caprella*, *Pycnogonum*, etc.), which are unique to this area, have a very poor ability to move.
2. The classes Mollusca, Echinodermata and Polypi, which are characteristic for this area, have a naked body.

NOTE. Of the areas of the Franco-Gallic and Norwegian coasts, this area corresponds exactly to the area with kelp.

List of all the animals in the area with nudibranchs.

§ 31. **Crustacea.**

Caprella linearis Kullen-Hellebæk. – *Phtisica marina* Hellebæk. – *Pycnogonum litorale* Mül. Kullen. – Snekkersten. – *Nymphon grossipes* Fabr.? Kullen. – *Anoplodactylus petiolatus* Kullen. – *Phoxichilidium femoratum* Rathke Kullen-Hellebæk. – *Phoxichilidium a new species*? Kullen.

Mollusca.

Doris pseudoargus Kullen. – *Doris* a new species? Related to *D. verrucosa*. Kullen. – *Cadlina laevis*[1] Kullen. *Polycera quadrilineata* Cuv. Kullen. *Okenia aspersa* Kullen.[2] *Hero formosa*[3] Hellebæk. – *Aeolidia papillosa* Cuv. Kullen-Hellebæk. – *Elysia viridis* Hven. – *Patella pellucida* L.

---

1. *Zool. Dan.*, vol. 47.
2. With elongated body, much taller than wide, orange, thinly curved tail, 3 simple lobes on both sides of the back, 6 tentacle-like appendages, of which two are twice as long, eight respiratory organs at the anus; 4–5 lines [≈ 1.016–1.27 mm] long, 2 lines [≈ 0.508 mm] wide.
3. Body oblong, anteriorly obtuse, posteriorly tapering, milky white with a light-yellow line across; two very large triangular lobes projecting from the margin of the anterior part of the body, bent backwards; of six pairs of respiratory organs, the second is twice as large as the first, the others gradually smaller.

Kullen. – *Tectura virginea* Mül. Kullen-Landskrona. – *Testudinalia testudinalis* – *Lepidochitona cinerea* – *Boreochiton ruber* L. Kullen-Landskrona. – *Halocynthia papillosa* Kullen[1]-Hven.

Echinodermata.

Psolus squamatus (*Holuthuria squamata Zool. Dan.*) Kullen-Landskrona – *Psolus phantapus* (*Holuthuria phantopus Zool. Dan.*) Kullen-Hellebæk. – *Cucumaria frondosa* (*Holuthuria pentactes Zool. Dan.*) Kullen-Landskrona. – *Thyone fusus Zool. Dan.*? Landskrona.

Polypi.

Metridium senile, Actinia plumosa Zool Dan., vol. 88 between Hven and Landskrona. – *Hormathia digitata, Actinia digitata Zool. Dan.*, vol. 133. Kullen. – *Stomphia coccinea, Actinia coccinea Zool. Dan.*, vol. 63 between Hven and Landskrona. – *Sagartiogeton viduatus Actinia viduata Zool. Dan.*, vol. 63 Kullen. – *Abietinaria abietina* Kullen. – *Hydrallmania falcata* together with the previous. – *Kirchenpaueria pinnata* together with the previous. – *Tubulipora liliacea* between *Laminaria* at Hven. – *Crisia eburnea* Lamour together with the previous – *Scrupocellaria* a new species?[2] found together with the previous species.

---

1. Three other species of this genus that I have found at Kullen are perhaps new.
2. It occupies a place between *Bugulina avicularia* and *Scrupocellaria scruposa*; it is closest to the latter in the form of the zooids, and to the former in its shape.

– *Flustra foliacea.* L. Kullen. – *Caberea ellisii* Kattegat. – *Electra pilosa* L. Hven. – *Alcyonidium gelatinosum* Johnst. Kullen. – *Corallina officinalis* L. [Rhodophyceae] Kullen-Hellebæk. – *Phymatolithon calcareum* [Rhodophyceae] – *Haliclona oculata, Dysidea fragilis,* and four other species which could be new.

The area with whelks.

(Marked in yellow on the Plate).

§ 32. **Depth.**
The limit of this area corresponds to the part of the Sound which, due to the nature of the bottom, is called the area of shells. It thus includes the entire lower part of the sea gorge, which forms the Sound between Scania and Zealand. This part of the Sound cannot therefore, in the stricter sense, be classified into regions defined solely by vertical distribution, but its narrow width gives it the characteristics of a distinct region. And in this area the animals are affected by other external factors than in the other areas. The bottom consists of dissolved shells and mud, which they can very easily penetrate. Therefore, the animals occur in burrows or various caves. For the most part, the anterior part of the body protrudes because the ferocious ones lurk in wait for swimming animals that pass by; they are all carnivores. As we have previously shown, there are no plants here. For this reason, this area differs greatly from the others, which facilitates such varied and

extensive vegetation; it is easy to see that this difference is of great importance in forming the basis of the characteristics of the animals. The water is therefore also different because it is calmer at the bottom; the waves cannot affect the bottom, and the intensity of the current is never as strong as it is in the adjacent area. Here, too, we do not rediscover the splendor and variety of colors which, together with the remarkable vegetation, aroused our admiration; the variety of forms and perhaps the amount of individuals in this area is very large. Of crustaceans, the decapods seem more than the others to be characteristic to this area; *Inachus phalangium*, *Hyas araneus* and *Pagurus bernhardus* are very abundant. Of polychaetes, *Aphrodita aculeata* is found only at very great depths, that is, around Hven; *Lepidonotus squamatus* mostly only in empty shells; *Scoletoma fragilis*, together with *Scoloplos armiger* on the seabed, corresponds to earthworms on land. In much the same way as with the previously mentioned species it is with *Nephtys ciliata* and *N. assimilis* and *Glycera alba*; *Nereis pelagica* is called the Worm King because of its size, speed and luster. The characteristic animals consist of molluscs. *Buccinum undatum* and *Neptunea antiqua* occur in large numbers on most of the bottom, so they are used by fishermen as bait.[1] *Aporrhais pespelecani*, *Turritella*

1. In order to catch them, a wicker basket, in which a dead cod (*Gadus*) is placed, is immersed on the clay bottom; with great ferocity they eat the fish for which they themselves are lovely food. Thus, when the wicker basket is filled with these animals, it is pulled up to the surface.

ungulina and *Antalis entalis* penetrate into the ocean depths through the clay layer. Of bivalves, special mention should be made of: *Aequipecten opercularis*, which also hides deep in the clay; for although there is a large quantity of its empty shells, one very rarely finds even a very small specimen of it containing a living animal; and *Nuculana minuta*, which occurs in large quantities, and penetrates the clay with very high speed and a long foot.[1] *Modiolus modiolus* is particularly frequent off Hellebæk, where there is clay not far from the coast. Therefore, it is praised when it quite often gets caught by the fishermen, due to the pleasant taste. *Acanthocardia echinata*, whose very long red foot strongly resembles a finger, hides in the clay, and avoids the dredge very easily; the same applies to *Arctica islandica*. Also to be mentioned: *Bosemprella incarnata*, whose shell, although very frequent, is mostly empty; *Hiatella arctica*, which with threads [byssus] attaches to the shells of others; and *Phaxas pellucidus*, through whose very thin shell the animal is visible.

Among the echinoderms, *Echinocardium flavescens*, *Solaster endeca* and *Crossaster papposus* occur only at very great depths; *Ophiopholis aculeata* and. *Amphiura fili-*

---

1. The food for these animals, which reside at such great depths, can best be seen in clay, which is picked up from the bottom, and is stored in large jars and poured over with water. Here they can live a long time, and a large number of small species that would otherwise evade our eyes appear.

formis live primarily in empty shells, to which they cling very tightly, so that they are torn to pieces when removed. Of polyps, there is special reason to draw attention to *Pennatula phosphorea* and *Virgularia mirabilis*, most of which protrude from the clay like a plant. I have seen few specimens of those with a slow motion penetrating the clay.

The general character of the animals in this area can be described in these few words: most of the animals that occur here are carnivorous and they live more or less under a cover of clay. If they have a shell, it is rarely of great thickness.

List of all animals in the area with whelks.

§ 33. **Crustacea.** *Inachus phalangium* Lamk. Hellebæk, Kullen. – *Hyas araneus* Leach. Kullen. – Landskrona. – *Pagurus bernhardus* Fabr. Kullen-Hven. – *Galathea strigosa* Fabr. Kullen, Hellebæk. – *Homarus gammarus* Kullen. – *Amphitoe*, a new species?[1] Kullen. – *Ericthonius punctatus* Hellebæk. – *Diastylis rathkei* Kullen-Hven. – *Leucon nasica* Kr. at the same locality. – *Diastylis lucifera*

1. With its long tentacles it makes a flat circular hole, slightly depressed, from the center of which the red head protrudes, but the rest of the body is hidden.

Kr. together with the previous. – *Idotea balthica*[1] Hellebæk. – *Balanus balanus* particularly among *Modiolus*.

Annulata.

Aphrodita aculeata Baster Kullen-Hven. – *Halithea hystrix* Aud. and Edw. Hornbæk Kröyer. – *Lepidonotus squamatus* Kullen-Hven. – *Lepidonatus laevis* Ørsd. Hven. – *Harmothoe assimilis* Ørsd. Hven. – *Scoletoma fragilis* Ørsd. Kullen-Hven. – *Nereis pelagica* L. – *Nereimyra punctata* Ørsd. Kullen-Hven. – *Syllis armillaris* Ørsd. Kullen-Hven. – *Eulalia viridis* Sav. The same locality. – *Eumida sanguinea* Ørsd. Hellebæk. – *Eteone flava* Ørsd. Hven. – *Eteone maculata* Ørsd. Kullen. – *Eteone pusilla* Ørsd. Hven. – *Phyllodoce groenlandica* Ørsd. Kullen. – *Nephtys ciliata* – *Nephtys assimilis* Ørsd. – *Glycera alba* Ørsd. – *Goniada maculata* Ørsd. Hellebæk-Hven – *Scoloplos armiger* Blainv. Kullen-Hven. – *Dipolydora coeca* Hven. – *Trochochaeta multisetosa* Ørsd. Hven. – *Cirratulus cirratus* Kullen-Hellebæk. – *Dodecaceria concharum* Ørsd. Hellebæk. – *Travisia forbesii* Kullen. – *Ophelina acuminata* Ørsd. Landskrona. – *Polyphysia crassa* Ørsd. Hven. – *Chaetopterus norvegicus* Sars? Hellebæk. – *Spirobranchus triqueter* – *Sabella pavonina* Kullen-Hven. – *Sabella* a new species? Hellebæk. – *Sabellaria* a new species? Kullen. – *Terebellides* a new species? Hven. – *Amphitrite cirrata* – *Amphicteis gunneri* Hellebæk. –

---

1. The only species of this genus found in the sea. Perhaps it should form its own genus.

Flabelligera affinis[1] Hellebæk. – *Pherusa plumosa* Blainv. – *Praxillella praetermissa*[2] Hellebæk. – *Galathowenia oculata* Hellebæk.[3] – *Arenicola marina* between Hven and Landskrona. – *Mesopachys marina* nob. Kullen.[4]

Apoda [should be "Vermes" by Ørsted].
Notoplana atomata Ørsd. Kullen-Hellebæk. – *Leptoplana nigripunctata* Ørsd. Kullen. – *Cryptocelides loveni* Hven. – *Cephalotrix linearis* Hven. – *Cephalothrix rufifrons* Ørsd. – *Oerstedia dorsalis* Kullen. – *Oerstedtia dorsalis* Kullen. – *Nemertes flaccida* Ørsd. – *Nemertes pusilla* Ørsd. the same locality. – *Nemertes maculata* Ørsd. Hellebæk. – *Polystemma roseum* Ørsd. – *Polystemma pellicudum* Ørsd. the same locality. – *Cerebratulus marginatus* Renieri?

1. *Ann. d. sc. Nat.*, vol. 6, pl. 1.
2. It occupies a position in between *Clymenia amphistoma* and *C. ebiensis* Aud. and Edw.; as for the shape of the head, it is closest to the latter, and as far as the anus is concerned, the former. It is characterized by 24 segments, 10 anterior and 3 posterior very short, the other 4–5 times as long as wide, hemispherical head, very small incisions at the anus.
3. General characteristics: very thin filamentous body consisting of numerous very distinct segments, the length of the segments exceeds the width greatly, club-shaped head, mouth terminal, two very small eyes, tail depressed, stiff bristles as in *Clymenia*. Differs from *Clymenia* which it is closest to, especially by the shape of the head, tail and pygidium. [Arwidsson, I. 1906: This is obviously *Myriochele oculata*].
4. Genus of the family Lumbricillus. Rod-shaped body consisting of 24–25 indistinct very short segments, no distinct head, parapodia ventral, 4 bundles of capillary setae in all segments. With twisted free esophagus, without any constriction, it differs from all genera of this family.

Hven. – *Malacobdella grossa* Blainv. (*Hirudo grossa* Z. D.) within *Arctica islandica*. – *Ichtyobdella sanguinea* nob. Hellebæk. – *Anguillula oculata* nob. Kullen.[1] – *Phascolosoma concharum* nob. (*Sipunculus* sp. Rathke in *Naturhistorie-Selskabets Skrifter*, vol. 5) within *Dentalium*, *Turritella ungulina* and others Kullen-Hven.[2]

Mollusca.

Gastropoda.

Velutina velutina Hellebæk. – *Buccinum undatum* L. – *Neptunea antiqua* – *Neptunea* (sp?) Kullen. – *Boreotrophon clathratus* Kullen-Hven. – *Pleurotomoides* sp. Kullen. – *Aporrhais pespelecani* – *Antalis entalis* Kullen-Hven.

Bivalvia.

Aequipecten opercularis Kullen-Hven. – *Palliolum striatum* [*Pecten striatus Zool. Dan.*, vol. 60] the same locality. – *Pseudamussium peslutrae* the same locality. – *Heteranomia squamula* L. [including its synonym *Anomia aculeata* Mül.] Kullen-Landskrona. – *Pododesmus patelliformis*? – *Anomia* a new species? All between Hven and Landskrona. – *Nucula* a new species? Hven. – *Nucula nucleus* Kullen-Hven. – *Nuculana pernula* – *Leda intermedia* nob.[3] [*Nuculana* sp.

---

1. Distinguished by its two brown eyes.
2. It particularly differs from *Phascolosoma granulatum* Leuchart (*Breves animalium descriptiones*, p. 5), which is very similar, in that the anterior part of the body is thinner and much longer.
3. It occupies a position between *L. rostrata* and *L. complanata* Mül [= *Nuculana minuta*] (Krøyers *Tidsskrift*, vol. 4, part 1, p. 90); with the former because it is of almost the same shape and furrowed in the same way, and with the latter because it is of the same size.

dub.] Hven. – *Modiolus modiolus* – *Musculus subpictus* – *Acanthocardia echinata* Kullen-Hven. – *Abra tenuis* Leach. Kullen-Hven. *Abra nitida* (*Mya nitida* Mül. with Beck's authority) – *Gari fervensis* Hellebæk-Landskrona. – *Thyasira flexuosa* Hven. – *Bosemprella incarnata* Kullen-Hven. – *Astarte sulcata* Hven. – *Astarte striata* Kullen. – *Lucinoma borealis* in Kattegat. – *Arctica islandica* Lamk. – *Chamelea gallina* L. Kullen-Hven. – *Phaxas pellucidus* Penn. the same locality. *Hiatella arctica* Lamk. – *Mya truncata* L. Hellebæk-Hven.

Echinodermata.

Echinocardium flavescens Zool. Dan. – *Echinocyamus pusillus* Mül Kullen. – *Crossaster papposus* Forbes (*Asterias papposa*) Hellebæk-Hven. – *Solaster endeca* Forbes (*Asterias endeca*) together with the previous. – *Astropecten irregularis* (*Asterias arantiaca Zool. Dan.*) Kullen-Hven. – *Stichastrella rosea* (*Asterias rosea*) Hven. – *Ophiopholis aculeata* (*Asterias aculeata*). – *Amphiura filiformis* (*Asterias filiformis*) Kullen-Hellebæk. – *Ophiura ophiura* (*Asterias ophiura*) – *Ophiothrix fragilis* M. T. (*Asterias fragilis*) where?

Polypi (Cnidaria, Entoprocta, Porifera, Bryoza). *Edwardsidae* sp. indet.[1] Kullen. – *Pennatula phosphorea* L. Kullen-Hven. – *Virgularia mirabilis* Lamk. together with the previous. – *Pedicellina cernua* on *Antalis* at Kullen. – *Barentsia gracilis* Sars together with the previous. – *Cliona celata* Grant. in between *Modiolus modiolus* – *Tethya* a new species? Hven. – *Alcyonium* a new species?

The horizontal distribution of the animals in the Sound.

§ 34. Above, we have seen that the salinity of the Sound slowly decreases from the northern to the southern part, causing a great difference of vegetation in the southern and northern parts. We will now show that the animals are affected in the same way, so that in the southern part of the Sound they largely correspond to the animals in fresh water. In the northern part there are in fact largely the same species as in the Kattegat; Arctic forms also occur here. I have previously shown[2] that among marine animals alone, there are nine species common to Greenland and the Sound. The island of Hven forms a clear

1. This very species-rich genus should certainly form the basis of a new order or at least a family forming a transition between Actiniae and Hydrae. As for the nature of Hydra's body and esophagus, they are completely similar; it is equipped with 16 tentacles in two rows, which, like with the Actiniae, can be pulled together. 3 lines [≈ 0.762 mm] long, 1 line [≈ 0.254 mm] wide. Sars means that it appears to be close to *Actinia prolifera*.
2. A.S. Ørsted, *Conspectus annnulatorum Danicorum*, p. 8.

boundary for them because it lies like a dam that resists the flow of water from the north. This seems to be the reason why a very large number of species gather in the Sound to the north and for the most part around Hven. Therefore, when we stay on this small island, we find a greater number of species than if we stay for a much longer time elsewhere. Kullen is perhaps the only place that can be compared to Hven in this respect. While *Antalis entalis*, *Aequipecten opercularis*, *Pseudamussium peslutrae* and *Palliolum striatum*, *Nucula nucleus*, *Thyasira flexuosa*, *Gari fervensis*, *Astarte sulcata*, *Arctica islandica*, *Travisia forbesii*, *Sabella pavonina*, *Pennatula phosphorea*, *Virgularia mirabilis* and several other species are encountered. However, the scarcity characteristic of the Baltic Sea already begins to appear half a mile to the north from Hven, so that the southern part of the Sound resembles the Baltic Sea. *Peringia* dominates here; to this are added species of genera that are characteristic to fresh water, such as *Lymnaea stagnalis* and *Eupaludestrina stagnorum*, as well as species that elsewhere are restricted to fresh water, such as *Bothrioplana semperi* and *Planaria torva*. In addition, a large number of insect larvae occur here, especially Diptera.

NOTE 1. Many species that are distributed over most of the bottom in much the same way occur in some places accumulated in such large numbers that it must almost be considered their special habitat. Although it is of great importance to know the habitats of all species, both the rare and the common ones, I will here only indicate the habitats of the species that are the most frequent in each place. Although I have found a few species represented by few or damaged individuals, I have no doubt that if they should be examined by adequate studies, there may be places where they are extremely abundant and numerous. These species are found in large numbers in the places listed on the first plate with letters. *Pagurus bernhardus* at Hellebæk (a). *Scoletoma fragilis* at Hven (b). *Nereis pelagica* at Hellebæk (c). *Hediste diversicolor* next to Clasens Have (d). *Nephtys ciliata* between Hellebæk and Viken (e). *Scoloplos armiger* between Viken and Höganäs (f). *Spio seticornis*, *Jaera albifrons*, *Corophium volutator* and *Monocelis lineata* occur mixed with each other on Kalvebod beach by Ny Badehus (g). *Procerodes littoralis* and *Lineus ruber* between Copenhagen and Trekroner (h). *Patella pellucida* at Kullaberg (i). *Turritella ungulina* and *Aporrhais pespelecani* between Gilleleje and Kullen (k). *Antalis entalis* between Sletten and the coast of Scania (l). *Aequipecten opercularis* between Hellebæk and Viken (m). *Nuculana minuta* at Hven (n). *Modiolus modiolus* at Hellebæk (o). *Musculus subpictus* at Hven (p). *Spisula* a new species? at Mölle (Kullen) (q). *Astarte sulcata* at Hven (r). *Arctica islandica* between Hven and Scania (s). *Mya arenaria* and *Cerastoderma edule* in the low-lying area Disken (t) and at Kalvebod beach (t). *Psolus phantapus* at Hellebæk and at Kullen (u). *Echinus esculentus* between Skovshoved and Scania (v). *Echinus esculentus* between Sletten and Hven (x). *Amphiura filiformis* at Kullen (y). *Virgularia mirabilis* and *Pennatula phosporea* at Kullen (ö)

NOTE 2. The number of animal species (with the exception of infusories [Ciliophora]) found so far in the Sound is 427. They are distributed between the different classes as follows:

Fish:	90
Crustaceans (Crustacea):	77
Spiders (Arachnidae):	3
Worms (Vermes):	110
Radiata:	58
Total:	338

§ 35. Above we have seen how the water influences the nature of the bottom, and moreover how the vegetation is affected by both the water and the bottom; we now easily understand how the distribution of animals over the different areas is conditioned not only by the properties of the water and the bottom, but also by the vegetation. How the animals are dependent on the plants, and the plants on external inorganic influences, we can only now clearly understand, as we have seen that when one part of these changes, the other also changes. The sandy area corresponds to the areas with brown algae, green algae and *Trochus* [trochoid snails]. In the same way, the rocky area and the areas with kelp, red algae and nudibranchs correspond to each other. Finally, the area with shells and the area without plants and whelks correspond to each other. If you have realized how they are interconnected, then the nature of the individual parts appears. So, if it is only known which animal is found at a depth of 11 fathoms in the Sound, the conditions under which it lives are easily concluded. We know that at this depth the bottom is covered with shells and completely without plants. Thus, we can understand in what way organisms are affected by external conditions that are of great importance to physiology. I postpone a more careful study of this matter to a later date, and I will just state that most lower sea creatures are not less affected by external conditions than plants, so that it seems possible to determine them in the same terms with which Unger has defined plants (bodenstäte, bodenholde, bodenwage [species that grow only on a single chemically specific soil type, species that prefer a particular soil type, though they also grow elsewhere, and generalists, respectively]). Here, however, it is a problem that the other external influences change together with the diversity of the bottom, so that it is often difficult to determine by which one the animals are particularly affected. Therefore, if an animal is not found except on a bottom covered with shells, the question is whether it is due to the conditions of the shells, or the light conditions that change by increasing depth, or in the water pressure, or in the great calm, and one realizes that the animals are very often affected by other factors than by the bottom itself. This table shows the internal connection between the areas.

Depth	0–50 feet	50–65 feet	65–? feet
Bottom covered by	sand	stones	clay
Plants	Green algae Brown algae *Zostera marina*	Red algae	absent
Mollusca	*Trochus* [trochoid snails]	Nudi- branchia	whelks

General results obtained from studies of the Sound.

§ 36.

1) The Sound originated either at the beginning of the present [Pleistocene] period or at the end of the Tertiary period.
2) If it was formed by a flood during the present [Pleistocene] period of the Earth, it is very likely that Zealand and Scania were connected by just one or two headlands.
3) If the coasts that surround the Sound were impacted by earthquakes after their formation, it appears that there was minor uplift of the coasts of Zealand and Scania.
4) It is not likely that the shape of the Sound has undergone major changes since it formed. On the coast of Zealand, it has been particularly affected by sedimentation, in the Scanian coast especially by erosion.
5) The nature of the Scanian coast between Glumslöv and Landskrona clearly shows that the special formation of plastic blue clay with cracks completely without fossils, which has been shown to exist both in Jutland (close to Fredericia) and on Funen and on Zealand by Forchhammer (Kröyers *Tidsskrift*, vol. 1, p. 209), also exists in Scania.

6) There are only three sediment types in the Sound: sand, clay and limestone shells (or rocks brought together); in all the places whose depth is greater than 8–10 fathoms, the bottom consists of clay, and where it is shallower, of sand. Calcareous shells or aggregations of rocks are only found where sand and clay have been washed away by the current.
7) If these conditions are not taken into account, the true composition of the variety of geological formations considered as normal cannot be understood.
8) Every geological formation consists of three parts, of which the first corresponds to sand, the second to clay and the third to calcareous shells.
9) Thus, during the formation of the Earth, nature has followed the same law that applies in the organic realms, an evolutionary development along parallel trajectories with structurally analogous parts.
10) In the Tertiary formation, the glacifluvial sand (Rullestenssandet) corresponds to the low-lying places in the Sound, and the glacifluvial clay (Rullestensleret) to the deeper places.
11) Thus, the two parts into which Denmark's Tertiary formation has been divided by Forchhammer were formed at the same time.
12) If Denmark had been inundated by a flood originating in the east, the bottom of the flooded places would have consisted of sand, not clay.
13) If the bottom consists of dispersed masses, due to

its nature, nothing can be concluded with certainty about the sediment layers that exist at the same depth.

14) The algae of the Sound are distributed over three areas, each of which has its own sub-areas, which are divided into the orders green algae, brown algae and red algae, such that red algae occur at the greatest depth, brown algae at intermediate depth and green algae at the shallowest depth.
15) This distribution is mainly due to the fact that the various light rays that penetrate the sea are refracted; when the red rays reach the greatest depth, there are red algae, thereafter orange colored, because the blue rays penetrate the least; Blue-green algae [Cyanobacteria] dominate at the sea's surface.
16) The striking observation that entire suborders of algae are defined by their shared pigmentation—or, put differently, that colour, which in most plants is a trait of minor taxonomic significance, here serves as a basis for higher-level classification—can be explained by the fact that algae, unlike other plants, are uniquely and variably influenced by light, upon which pigmentation ultimately depends.
17) Also, in the area of the algae, evolution takes place along parallel lineages with analogous structural features.

18) Most of the lower marine animals are thus affected by external factors, such that they, like the plants, are linked to specific habitats.
19) Consequently, marine animals can be attributed to certain habitats.
20) Molluscs (Mollusca) provide the best basis for a division of habitats.
21) According to the distribution of the molluscs (Mollusca), three habitats can be identified: one with *Trochus* [trochoid snails], one with nudibranchs and one with whelks.
22) These areas have their basis partly in different vegetation and partly in the difference in external factors, which are due to different depths.

Explanation of the woodcuts and plates.

1. p. 177. The Ball's dredge.
2. p. 191. Cross-section of the mound at Villingebæk. a) layer of loose sand. b) layer with marine organisms.
3. p. 193. Layer that occurs close to Landskrona after a channel has formed. a) soil with sand, b) layers of marine animals, c) layers of clay mixed with oxidized iron, d) layers of stones, e) plain yellow clay.
4. Steep slopes between Sletten and Humlebæk. a) plain yellow clay, b) layer with very soft clay without stones, c) layer with stones, d) conglomerate with chalk.
5. Steep slopes at Glumslöv in Scania. a) plain yellow clay

mixed with stones, b) soil layers, c) layers with fresh-water chalk, d) blue clay that is plastic and can be split.

Plates I–II.
Figures 1–6. The cross-sections of the Sound, which explain the different depths of different parts of the Sound. The numbers added indicate the consecutive depths. On plate I the yellow color indicates clay and the area with whelks, the red color indicates areas with red algae, the dark brown indicates areas with kelp; both indicate an area with stones and nudibranchs; the light brown color indicates areas with brown algae, the green color an area with green algae, the blue color an area with blue-green algae [Cyanobacteria]; these three colors also simultaneously indicate areas with *Trochus* [trochoid snails] and a sandy area. Plate II shows how these areas can be considered a continuation of the rocky zones. The different colors indicate the same as on plate I.

Corrections.

It is:	It should be:
P. 13 6 v. im Iul.	Iun.
P. 32 12 v. membronum	membrorum.
P. 32 13 v. formatiorum	formationum.
P. 41 8 v. *Conferva distans* Dillv.	*Conferva diffusa* Dill. ?

Additions.

§ 12 Note. It is stated that Tertiary clay is completely without fossils; but I was wrong in that the layers of the Tertiary formation which had been discovered by Forchhammer on the islands of Langeland, Ærø, Als and others (see *Proceedings and Papers of the Royal Danish Academy of Sciences and Letters* (1842), no. 5) belong to this middle part of the Tertiary formation. By this observation my view has been miraculously confirmed, for *Arctica islandica* turns out[1] to dominate in that part of the Tertiary clay, and it is already known that *Cerastoderma edule*, *Mytilus edulis* and *Tritia reticulata* are dominant in Tertiary sand; but in the sea at the depth of the [Pleistocene] period, viz. clay, *Arctica islandica* is characteristic of deeper areas with clayey substrates, while *Cerastoderma edule* and others are characteristic of shallower areas with sandy substrates, such that there is a close analogy between the present and the Tertiary seas.

1. The word nom in the original text is a typographical error for nam.

Ørsted, A.S. 1844. *De Regionibus Marinis.*

List of taxa mentioned in the text

The higher taxa are listed in the order they usually appear in standard taxonomic publications. However, genera and species are listed in alphabetical order under each higher taxon. The reason is that some of these names represent taxa of uncertain identity, and a different order would therefore have been impossible to decide upon.

A few terms used by Ørsted are not clearly defined. When he used the name "Conferva" (pp. 38–39) he probably meant green algae (Chlorophyta). The taxa "Apoda" and "Vermes" (pp. 68, 79, 84) in his time usually contained various unrelated worm-shaped organisms. Ørsted mentioned some species in different grammatical forms in different parts of his text. Such varieties have not been included in this checklist.

Authorities for species and synonyms given by Ørsted have not been included in the list below but can easily be found in the original text.

Some of the names of taxa below are called "uncertain" in parentheses following the name. This term, used by the database WoRMS, covers the terms *nomen dubium* and *taxon inquirendum*. *Nomen dubium* is a name of uncertain application while *taxon inquirendum* is a name that should be investigated.

Organisms are synonymized by the use of the databases WoRMS and AlgaeBase.

Bacteria.

***Leucothrix mucor* Ørsted**
Ø1844: 44: *Leucothrix mucor.*

Cyanobacteria

This group was called Algae caeruleoviridis (p. 39) and Algae viridicaerulescens (p. 56).

***Blennothrix glutinosa* (Gomont ex Gomont) Anagnostidis & Komárek, 2001**
Ø1844: 39, 43: *Lyngbyea glutinosa.*

***Calothrix fasciculata* C. Agardh ex Bornet & Flahault, 1886**
Ø1844: 39, 43: *Callothrix fasciculata.*
Lange (1887) gives the same reference to Plate 2517, 1 for this and next species.

***Calothrix scopulorum* C. Agardh ex Bornet & Flahault, 1886**
Ø1844: 43: *Callothrix scopulorum.*
Lange (1887) gives the same reference to Plate 2517, 1 for this and previous species.

***Heteroscytonema crispum* (Bornet ex De Toni) McGregor & Sendall, 2018**
Ø1844: 43: *Lyngbyea crispa.*

***Isactis plana* Thuret ex Bornet & Flahault, 1886**
Ø1844: 43: *Rivularia (Scytochloria) plana.*

***Lyngbya aestuarii f. ferruginea* Gomont, 1892**
Ø1844: 43: *Lyngbyea ferruginea.*

***Lyngbya aestuarii* Liebman ex Gomont, 1892**
Ø1844: 39, 43: *Lyngbyea aestuarii*; 43: *Oscillatoria aestuarii.*

***Lyngbya lutea* Gomont ex Gomont, 1892**
Ø1844: 44: *Oscillatoria lutea.*

***Lyngbya majuscula* Harvey ex Gomont, 1892**
Ø1844: 43: *Oscillatoria maiuscula.*

***Lyngbyea lutescens* Liebmann (1838–1839)** (uncertain)
Ø1844: 39, 43, 47: *Lyngbyea lutescens.*
Figure from Liebmann, F.M. (1839: pl. VI, fig 5).

***Merismopedia litoralis* (Ørsted) Rabenhorst, 1865**
Ø1844: *Erythroconis littoralis.*

***Microcoleus chthonoplastes* Thuret ex Gomont, 1892** (uncertain)
Ø1844: 39: *Microcoleus chthonoplastes.*

***Lyngbya* C. Agardh ex Gomont, 1892**
Ø1844: 38, 39, 43, 56, 87: *Oscillatoria.*

***Rivularia atra* Roth ex Bornet & Flahault, 1886**
Ø1844: 43: *Rivularia atra.*

***Rivularia nitida* C. Agardh ex Bornet & Flahault, 1886**
Ø1844: 43, 47: *Rivularia nitida.*

***Rivularia pellucida* C. Agardh, 1824** (uncertain)
Ø1844: 43: *Rivularia pellucida.*
Drouet (1973: 166, 183) designated a type specimen from Landskrona, Sweden.

***Scytonema* C. Agardh ex Bornet & Flahault, 1886**
Ø1844: 42: *Scytonema.*

***Microcoleus fuscus* Ørsted, 1844**
Ø1844: 39, 43: *Microcoleus fuscus.*

***Spirulina subsalsa* Ørsted ex Gomont, 1892**
Ø1844: 39, 43: *Spirulina subsalsa.*

Rhodophyceae

Ø1844: 50, 51, 52, 56, 57, 59, 60, 71, 85, 87: Algae Rhodospermeae; 52, 56: Algae purpureae; 37, 59: Algae Florideae.

***Ahnfeltia plicata* (Hudson) Fries, 1836**
Ø1844: 47, 52: *Gigartina plicata.*

***Antithamnionella floccosa* (Müller) Whittick, 1980**
Ø1844: 43: *Conferva floccosa.*

***Callithamnion* Lyngbye, 1819**
Ø1844: 51: *Callithamnion.*

***Callithamnion tetragonum* (Withering) Gray, 1821**
Ø1844: 53: *Callithamnium fruticulosum.*

***Carradoriella elongata* (Hudson) Savoie & Saunders, 2019**
Ø1844: 53: *Ceramium brachygonium, Ceramium elongatum, Ceramium elongatum proliferum, Ceramium elongatum* var *denudatum.*

***Ceramium* Roth, 1797**
Ø1844: 38, 51, 53, 59: *Ceramium.*

***Ceramium ciliatum* (J. Ellis) Ducluzeau, 1806**
Ø1844: 53: *Ceramium ciliatum.*

***Ceramium diaphanum* (Lightfoot) Roth, 1806**
Ø1844: 40, 53: *Ceramium diaphanum.*

***Ceramium siliquosum var. elegans* (Roth) Furnari, 1999**
Ø1844: 40: *Conferva elegans.*

***Ceramium virgatum* Roth, 1797**
Ø1844: 53: *Ceramium rubrum.*

***Chondrus* Stackhouse, 1797**
Ø1844: 37, 38, 47: *Chondrus.*

***Chondrus crispus* Stackhouse, 1797**
Ø1844: 52: *Chondrus crispus; Chondrus crispus* var. *incurvatus.*

***Coccotylus brodiei* (Turner) Kützing, 1843**
Ø1844: 52: *Chondrus Brodiaei lingulatus.*

***Coccotylus truncatus* (Pallas) Wynne & Heine, 1992**
Ø1844: 52: *Chondrus Brodiaei angustissimus.*

***Colaconema daviesii* (Dillwyn) Stegenga, 1985**
Ø1844: *Callithamnium lanuginosum.*

***Compsothamnion thuioides* (Smith) Nägeli, 1862**
Ø1844: *Callithamnium thyoides.*

***Corallina* Linnaeus, 1758**
Ø1844: 66: *Corallina.*

***Corallina officinalis* Linnaeus, 1758**
Ø1844: 61, 74: *Corallina officinalis.*

***Cruoria pellita* (Lyngbye) Fries, 1835**
Ø1844: 40, 47, 50: *Cruoria pellita;* 50: *Erythroclathrus pellitus.*

***Cystoclonium purpureum* (Hudson) Batters, 1902**
Ø1844: 52: *Gigartina purpurascens.*

***Delesseria* J.V. Lamouroux, 1813**
Ø1844: 37, 38, 72: *Delesseria;* 51: *Dellesseria.*

***Delesseria sanguinea* (Hudson) J. V. Lamouroux, 1813**
See also *Odonthalia dentata* below.
Ø1844: 52: *Delesseria sanguinea.*

***Dilsea carnosa* (Schmidel) Kuntze, 1898**
Ø1844: 51, 52: *Iridea edulis;* 52: *Halymenia edulis.*

***Dumontia contorta* (S.G. Gmelin) Ruprecht, 1850**
Ø1844: 47, 52: *Dumontia filiformis;* 52: *Gastridium filiforme.*

***Erythrotrichia carnea* (Dillwyn) J. Agardh, 1883**
Ø1844: 42: *Ceramicola rubra;* 42: *Conferva ceramicola.*

***Furcellaria lumbricalis* (Hudson) J. V. Lamouroux, 1813**
Ø1844: 47, 52: *Furcellaria fastigiata.*

***Gaillona hookeri* (Dillwyn) Athanasiadis, 2016**
Ø1844: 53: *Callithamnium pyramidatum.*

Gastrocarpeae
Ø1844: 52: Gastrocarpeae
Probably Dumontiaceae.

***Gigartina* Stackhouse, 1809**
Ø1844: 51: *Gigartinae* species.

***Gloiocladia* J. Agardh, 1842?**
Ø1844: 52: *Gloiocladea.*

***Gloiocladia repens* (C. Agardh) N. Sánchez & Rodríguez-Prieto, 2007**
Ø1844: 52: *Sphaerococcus rubens.* It could also be *Chylocladia rigens* (C. Agardh) J. Agardh, 1851.

***Gracilariopsis longissima* (Gmelin) Steentoft, Irvine & Farnham 1995**
Ø1844: 52: *Gigartina confervoides.*

***Halymenia* C. Agardh, 1817**
Ø1844: 59: *Halymeniea.*

***Leptosiphonia brodiei* (Dillwyn) A.M. Savoie & G.W. Saunders, 2019**
Ø1844: 53: *Hutchinsia Brodiaei;* 53: *Hutchinsia penicillata.*

***Leptosiphonia fibrillosa* (C. Agardh) A.M. Savoie & G.W. Saunders, 2019**
Ø1844: 53: *Hutchinsia divaricata;* 53: *Hutchinsia tenuis.*

***Membranoptera alata* (Hudson) Stackhouse, 1809**
Ø1844: 52: *Delesseria alata.*

***Nemalion multifidum* (Lyngbye) Chauvin, 1842**
Ø1844: 47, 52: *Helminthora multifida;* 52: *Chordaria multifida.*

***Odonthalia dentata* (Linnaeus) Lyngbye, 1819**
Ø1844: 51, 52: *Odonthalia dentata.*

***Odonthalia dentata* (Linnaeus) Lyngbye, 1819**
Ø1844: 72: *Odonthalia sanguineis. Odonthalia dentata* exists within the region, but this is probably a error for *Delesseria sanguinea* (Hudson) J.V. Lamouroux, 1813.

***Palmaria palmata* (Linnaeus) F. Weber & D. Mohr, 1805**
Ø1844: 52: *Rhodomenia palmata.*

***Phycodrys rubens* (Linnaeus) Batters, 1902**
Ø1844: 52: *Delesseria sinuosa.*

***Phyllophora pseudoceranoïdes* (S. G. Gmelin) Newroth & A. R. A. Taylor ex P. S. Dixon & L. M. Irvine, 1977**
Ø1844: 52: *Chondrus membranifolius.*

***Phymatolithon calcareum* (Pallas) W. H. Adey & D. L. McKibbin ex Woelkering & L. M. Irvine, 1986**
Ø1844: 74: *Nullipora polymorpha.*

Polyidaceae Kylin, 1956
Ø1844: 52: Spongiocarpeae.

***Polyides rotunda* (Hudson) Gaillon, 1828**
Ø1844: 47, 52: *Polyides rotundus.*

***Polysiphonia* Greville, 1823**
Ø1844: 51: *Hutchinsia. Hutchinsia* is probably *Polysiphonia.*

***Polysiphonia stricta* (Mertens ex Dillwyn) Greville, 1824**
Ø1844: 53: *Hutchinsia roseola.*

***Polysiphonia urceolata f. lepadicola* (Lyngbye) Segi, 1951**
Ø1844: 53: *Hutchinsia lepadicola.*

***Ptilota* C. Agardh, 1817**
Ø1844: 72: *Ptilota.*

***Ptilota gunneri* P.C. Silva, Maggs & L.M. Irvine, 1993**
Ø1844: 52: *Pti[l]ota plumosa.*

***Rhodochorton purpureum* (Lightfoot) Rosenvinge, 1900**
Ø1844: 53: *Callithamnium Rothii.*

***Rhodomela confervoides* (Hudson) P.C. Silva, 1952**
Ø1844: 47, 52: *Rhodomela subfusca.*

***Spermothamnion repens* (Dillwyn) Magnus, 1873**
Ø1844: 53: *Callithamnium repens.*

***Stilophora nodulosa* (C. Agardh) P.C. Silva, 1996**
Ø1844: 50: *Ceramium tuberculosum.*

***Vertebrata fucoides* (Hudson) Kuntze, 1891**
Ø1844: 40, 53: *Hutchinsia nigrescens*; 53: *Hutchinsia violacea.*

Bacillariophyceae

Ø1844: 44, 59, 60: Diatomaceae.

***Achnanthes brevipes* C. Agardh, 1824**
Ø1844: 45: *Achnanthes brevipes.*

***Achnanthes intermedia* Kützing, 1833** (uncertain)
Ø1844: 45: *Achnanthes subsessilis.*

***Achnanthes longipes* C. Agardh, 1824**
Ø1844: 45: *Achnanthes longipes.*

***Achnanthes minutissima* Kützing, 1833** (uncertain)
Ø1844: 45: *Achnanthes minutissima.*

***Amphiprora constricta* Ehrenberg, 1843** (uncertain)
Ø1844: 44: *Amphiprora constricta.*

***Bacillaria adriatica* Lobarzewski, 1840** (uncertain)
Ø1844: 45: *Bacillaria adriatica.*

***Bacillaria elongata* Ehrenberg, 1838** (uncertain)
Ø1844: 45: *Bacillaria elongata.*

Bacillaria flocculosa **(Roth) Leiblein, 1827** (uncertain)
Ø1844: 45: *Bacillaria flocculosa.*

Berkeleya rutilans **(Trentepohl ex Roth) Grunow, 1880**
Ø1844: 45: *Bangia rutilans*; 45: *Schizonema rutilans.*

Ceratoneis closterium **Ehrenberg, 1839** (uncertain)
Ø1844: 44: *Ceratoneis Closterium.*

Cocconeis scutellum **Ehrenberg, 1838**
Ø1844: *Cocconeis Scutellum.*

Grammatophora marina **(Lyngbye) Kützing, 1844**
Ø1844: 45: *Diatoma marinum.*

Diatoma signata **Dujardin, 1842** (uncertain)
Ø1844: 45: *Diatoma signata.*

Diatoma vulgaris **Bory de Saint-Vincent, 1824**
Ø1844: 45: *Bacillaria vulgaris.*

Eunotia turgida **(Ehrenberg) Ehrenberg, 1837** (uncertain)
Ø1844: 44: *Eunotia turgida.*

Fragilaria lineata **(Dillwyn) Lyngbye** (uncertain)
Ø1844: 44: *Fragillaria lineata.*

Fragilaria nummuloides **(Dillwyn) Lyngbye, 1819** (uncertain)
Ø1844: 44: *Fragillaria numuloides.*

Frustulia obtusa **(Lyngbye) C. Agardh, 1824** (uncertain)
Ø1844: 45: *Echinella obtusa.*

Gaillonella moniliformis **Bory in Ehrenberg, 1838**
Ø1844: 44, 47: *Gallionella moniliformis.*

Grammatophora mexicana **Ehrenberg, 1840** (uncertain)
Ø1844: 45: *Grammatophora Mexicana.*

Himantidium faba **Ehrenberg, 1854** (uncertain)
Ø1844: *Eunotia Faba.*

Homalodiscus ovalis **Ørsted, 1844** (uncertain)
Ø1844: 44: *Homalodiscus ovalis.*

Homalodiscus vulgaris **Ørsted, 1844** (uncertain)
Ø1844: 45: *Homalodiscus vulgaris.*

Licmophora abbreviata **C. Agardh, 1831**
Ø1844: *Podosphenia abbreviata.*

Licmophora gracilis **(Ehrenberg) Grunow, 1867**
Ø1844: 45: *Podosphenia gracilis.*

Melosira lineata **(Dillwyn) C. Agardh, 1824** (uncertain)
Ø1844: *Gallionella lineata.*

Tessella interrupta **Ehrenberg, 1838** (uncertain)
Ø1844: 45: *Tessella interrupta.*

Xanthophyceae.

***Tribonema bombycinum* (C. Agardh) Derbès & Solier, 1851**
Ø1844: 41: *Bombycina stagnalis*; 41: *Conferva bombycina*; 41: *Conferva bombycina submarina*.

Phaeophyceae

Ø1844: 37: Algae Olivaceae; 38, 46, 47, 56, 57, 59, 60, 85, 87: Algae Melanospermeae.

***Chaetopteris plumosa* (Lyngbye) Kützing, 1843**
Ø1844: *Sphacelaria plumosa*.

***Chorda filum* (Linnaeus) Stackhouse, 1797**
Ø1844: 47, 48: *Chorda Filum*.

***Chordaria* C. Agardh, 1817**
Ø1844: 37, 50, 59: *Chordaria*.

***Chordaria flagelliformis* (Müller) C. Agardh, 1817**
Ø1844: 47, 49, 50: *Chordaria flagelliformis*.

***Cladostephus spongiosus f. verticillatus* (Lightfoot) Prud'homme van Reine, 1972**
Ø1844: 49: *Cladostephus verticillatus*.

***Desmarestia aculeata* (Linnaeus) Lamouroux, 1813**
Ø1844: 47, 48: *Desmarestia aculeata*; 48: *Desmarestia aculeata* var. *Plumosa*; 48: *Ectocarpus densus*.

***Desmarestia aculeata var. complanata* (C. Agardh) Ørsted, 1844**
Ø1844: *Desmarestia aculeata* var. *Complanata*.

***Desmarestia viridis* (O.F. Müller) J.V. Lamouroux, 1813**
Ø1844: 47, 48: *Dichloria viridis*.

***Dictyosiphon foeniculaceus* (Hudson) Greville, 1830**
Ø1844: 49: *Dictyosiphon foeniculaceus*; 52: *Gigartina plicata* var. *hippuroides*; 52: *Scytosiphon hippuroides*.

***Dictyota* J.V. Lamouroux, 1809**
Ø1844: 37, 48, 59: *Dictyota*.

Ectocarpaceae C. Agardh, 1828
Ø1844: 49, 59: Ectocarpeae.

***Ectocarpus siliculosus* (Dillwyn) Lyngbye, 1819**
Ø1844: 40, 47, 49: *Ectocarpus siliculosus*.

***Elachista fucicola* (Velley) Areschoug, 1842**
Ø1844: 47, 50: *Elachista globosa*; 50: *Elachista fucicola*.

***Elachista globulosa* (C. Agardh) J. Agardh, 1848**
Ø1844: 40: *Conferva globosa*.

***Elachista stellaris* Areschoug, 1842**
Ø1844: 50: *Elachista stellaris*.

Eudesme virescens **(Carmichael ex Berkeley) J. Agardh, 1882**
Ø1844: 52: *Mesogloiia Zosterae*; 52: *Aegira Zosterae.*

Fucaceae Adanson, 1763
Ø1844: 38, 46, 48, 59, 64: Fucoideae.

Fucus **Linnaeus, 1753**
Ø1844: 37, 61: 1; 48: Fucus varietas tenuior [Thinner varieties of *Fucus*].

Fucus serratus **Linnaeus, 1753**
Ø1844: 46, 48: *Fucus serratus.*

Fucus vesiculosus **Linnaeus, 1753**
Ø1844: 46, 48: *Fucus vesiculosus.*

Halidrys siliquosa **(Linnaeus) Lyngbye, 1819**
Ø1844: 47, 48: *Halidrys siliquosa.*

Laminaria **J.V. Lamouroux, 1813**
Ø1844: 38, 47, 48, 51, 57, 59, 61, 63, 71, 72, 74: *Laminaria.*

Laminaria digitata **(Hudson) Lamouroux, 1813**
Ø1844: 48: *Laminaria digitata*; 48: *Laminaria latifolia.*

Leathesia marina **(Lyngbye) Decaisne, 1842**
Ø1844: 47, 50: *Clavatella difformis*; 50: *Corynephora marina.*

Mesogloia vermiculata **(Smith) Gray, 1821**
Ø1844: 47, 52: *Mesogloia vermicularis.*

Petalonia fascia **(Müller) Kuntze, 1898**
Ø1844: 40, 47, 48: *Ilea fascia*; 48: *Laminaria fascia*; 48, 49: *Laminaria Fascia* var. *tenuior*; 47: *Ilia fascia* var. *Tenuior* [Thinner variety of *Petalonia fascia*].

Punctaria tenuissima **(C. Agardh) Greville, 1830**
Ø1844: 49: *Punctaria undulata.*

Pylaiella littoralis **(Linnaeus) Kjellman, 1872**
Ø1844: 40, 49: *Ectocarpus littoralis*; 49: *Ectocarpus brachiatus*; 49: *Ectocarpus compactus.*

Ralfsia fungiformis **(Gunnerus) Setchell & Gardner, 1924**
Ø1844: 49: *Zonaria deusta.*

Saccharina latissima **(Linnaeus) Lane, Mayes, Druehl & Saunders, 2006**
Ø1844: 48: *Laminaria saccharina.*

Scytosiphon lomentaria **(Lyngbye) Link, 1833**
Ø1844: 40, 47, 48: *Chorda lomentaria.*

Spermatochnus paradoxus **(Roth) Kützing, 1843**
Ø1844: 50: *Chordiariae paradoxa.*

Sphacelaria **Lyngbye, 1818**
Ø1844: 37: *Sphacellaria.*

***Sphacelaria cirrosa* (Roth) C. Agardh, 1824**
Ø1844: 49: *Sphacelaria cirrhosa.*

***Spongonema tomentosum* (Hudson) Kützing, 1849**
Ø1844: *Ectocarpus tomentosus.*

Sporochnaceae Greville, 1830
Ø1844: 48, 59: Sporochnoideae.

***Stilophora nodulosa* (C. Agardh) Silva, 1996**
Ø1844: 47, 50: *Chordaria tuberculosa*; 50: *Chaetophora nodulosa.*

***Stilophora tenella* (Esper) Silva, 1996**
Ø1844: 47: *Sporochnus rhizodes*; 50: *Chaetophora rhizodes*; 50: *Chordiaria rhizodes.*

***Zonarina liebmannii* Ørsted, 1844**
Ø1844: 47, 49: *Zonarina Liebmanni.*

Chlorophyceae

Ø1844: 59: Cor[n]fervoideae [Filamentous green algae]; 38, 39, 51, 56, 57, 59, 60, 85, 87: Algae Chlorospermeae; 39, 56, 58: Algae viridis; 36, 39: Algae Zoospermeae; 37: Conferva.

***Acrosiphonia arcta* (Dillwyn) Gain, 1912**
Ø1844: 40: *Conferva centralis*; 41: *Conferva arcta.*

Bryopsidales
Ø1844: 43: Siphoneae.

***Bryopsis plumosa* (Hudson) C. Agardh, 1823**
Ø1844: 43: *Bryopsis arbuscula.*

***Chaetomorpha ligustica* (Kützing) Kützing, 1849**
Ø1844: 39, 41: *Conferva perreptans.*

***Chaetomorpha linum* (Müller) Kützing, 1845**
Ø1844: 37, 40, 41: *Conferva aerea*; 39, 41: *Conferva Linnum*; 41: *Conferva rigida.*

***Chaetomorpha melagonium* (Weber & Mohr) Kützing, 1845**
Ø1844: 41: *Conferva Melagonium.*

***Cladophora flexuosa* (Müller) Kützing, 1843**
Ø1844: 40, 41: *Conferva gracilis*; 41: *Conferva elegans.*

***Cladophora fracta* (Müller ex Vahl) Kützing, 1843**
Ø1844: 39: *Conferva fracta.*

***Cladophora glomerata* (Linnaeus) Kützing, 1843**
Ø1844: 40, 41: *Conferva glomerata.*

***Cladophora hutchinsiae* (Dillwyn) Kützing, 1845**
Ø1844: 40, 41, 90: *Conferva distans.*

***Cladophora rupestris* (Linnaeus) Kützing, 1843**
Ø1844: 37, 40, 41: *Conferva rupestris.*

***Cladophora sericea* (Hudson) Kützing, 1843**
Ø1844: *Conferva sericea.*

***Conferva bombycina var. submarina* C. Agardh** (species dubium)
Ø1844: 41: *Bombycina marina.*

***Conferva hofmannii* C. Agardh** (uncertain)
Ø1844: 41: *Conferva Hofmanni.*

***Conferva hormoides* Lyngbye**
Ø1844: 41: *Conferva hormoides.*

***Conferva obtusangula* Lyngbye**
Ø1844: 41: *Conferva obtusangula.*

***Solenia* C. Agardh, 1824** (uncertain)
Ø1844: 43: *Solenia clathrata confervoidea.*
This is probably *Enteromorpha.*

***Spongomorpha aeruginosa* (Linnaeus) Hoek, 1963**
Ø1844: 40, 41: *Conferva uncialis.*

***Spongomorpha arcta* (Dillwyn) Kützing, 1849**
Ø1844: 40, 41: *Conferva vaucheriaeformis.*

***Ulothrix* Kützing, 1833**
Ø1844: 42: *Myxonema.*

***Ulothrix contorta* (C. Agardh) Trevisan** (uncertain)
Ø1844: 40: *Ulothrix contorta.*

***Ulothrix flacca* (Dillwyn) Thuret, 1863**
Ø1844: 41: *Hormiscia flacca;* 41: *Myxonema flaccum;* 42: *Conferva flacca.*

***Ulothrix floccosa* (Vaucher) Ørsted**
Ø1844: 39: *Ulothrix floccosa;* 43: *Ulothrix (Bispora) floccosa.*

***Ulothrix subsalsa* Ørsted (uncertain)**
Ø1844: 42: *Ulothrix subsalsa.*

Ulvaceae J.V. Lamouroux ex Dumortier, 1822
Ø1844: 37, 40, 43, 59: Ulvaceae.

***Ulva* Linnaeus, 1753**
Ø1844: 37, 38, 39, 40, 64: *Ulva.*

***Ulva clathrata* (Roth) C. Agardh, 1811**
Ø1844: 39: *Ulva clathrata;* 43: *Enteromorpha clathrata.*

***Ulva compressa* Linnaeus, 1753**
Ø1844: 40, 43: *Ulva compressa.*

***Ulva flexuosa subsp. paradoxa* (C. Agardh) Wynne, 2005**
Ø1844: 43: *Enteromorpha (Scytosiphon) erecta.*

***Ulva lactuca* Linnaeus, 1753**
Ø1844: 40, 43, 66: *Ulva lactuca.*

***Ulva lactuca var. latissima* (C. Linnaeus) T. Edmondston**
Ø1844: 40, 43: *Ulva latissima.*

***Ulva linza* Linnaeus, 1753**
Ø1844: 40, 43: *Ulva Linza.*

***Urospora penicilliformis* (Roth) Areschoug, 1866**
Ø1844: 40,41, 42, 47: *Hormiscia penicilliformis;* 40, 41, 42: *Hormiscia assimilis.*
Synonymy according to Lokhorst & Trask (1981).

Viridiplantae

Ø1844: 85: Plantae.

Characeae Gray, 1821
Ø1844: 60: Characeae.

***Najas* Linnaeus**
Ø1844: 60: *Naias.*

***Ruppia maritima* Linnaeus, 1753**
Ø1844: 60: *Ruppia maritima;*
60: *Ruppia rostellata.*

***Tolypella nidifica* (Müller) Leonhardi, 1864**
Ø1844: 60: *Chara nidifica.*

***Zannichellia palustris* Linnaeus**
Ø1844: 60: *Zanichellia palustris.*

***Zostera marina* Linnaeus, 1753**
Ø1844: 43, 46, 49, 55, 60, 61, 64, 65, 66, 71, 85: *Zostera marina.*

***Zygnema littoreum* Lyngbye** (uncertain)
Ø1844: 41: *Zygnema littoreum.*

Fungi ("Lichenes")

Ø1844: 47, 48, 60: Lichineae.

***Agonium centrale* Ørsted, 1844.**
Ø1844: 44, 47: *Agonium centrale.*
Heiberg (1863) gives a comment on this species based on drawings given to him by Ørsted and concluded that it is not a diatom but rather a kind of oscillatorian.

***Lichina* C. Agardh, 1817**
Ø1844: 37: *Lichina.*

***Lichina confinis* (Müller) C. Agardh, 1821**
Ø1844: 40, 47, 48: *Lichina confinis.*

***Verrucaria maura* Wahlenberg, 1803**
Ø1844: 47, 60: *Verrucaria Maura.*

Ciliophora

Ø1844: 84: Infusoria.

Porifera

Ø1844: 61: Spongia.

***Cliona celata* Grant, 1826**
Ø1844: 75: *Cliona celata.*

***Dysidea fragilis* (Montagu, 1814)**
Ø1844: 75: *Spongia fragilis.*

***Haliclona (Haliclona) oculata* (Linnaeus, 1759)**
Ø1844: 75: *Spongia coalita;*
Spongia ossiformis Müller, 1776 (taxon inquirendum).
The shape of this species called *Spongia ossiformis* on plate XL in Müller (1788) represents a growth form where branching has just begun. The species is common in the Sound (Køie & Kristiansen 2014).

***Tethya* Lamarck, 1815**
Ø1844: 82: *Tethya* nov. sp.?

Radiata

Ø1844 84: Radiata.
This is a polyphyletic grouping of unrelated taxa, Coelenterata + Echinodermata (Cuvier 1817) used in the time of Ørsted.

Cnidaria, Anthozoa

Ø1844: 61, 72: Actiniae; 71, 74, 81: Polypi.

***Alcyonium* Linnaeus, 1758**
Ø1844: 82: *Acyonium* nov. sp? Probably misspelling for *Alcyonium*.

***Alcyonium digitatum* Linnaeus, 1758**
Ø1844: 71: *Lobularia digitata*.

Edwardsiidae
Ø1844: 81: *Hydropsis gelatinosa*. *Hydropsis gelatinosa* is close to *Gonactinia prolifera* according to Sars. *Edwardsia pallida*, *E. longicornis* and *Paraedwardsia arenaria* usually have 16 tentacles. *E. longicornis* is found in the Sound (Carlgren 1945).

***Hormathia digitata* (Müller, 1776)**
Ø1844: 74: *Actinia digitata*; 74: *Isacmaea digitata*.

***Metridium senile* (Linnaeus, 1761)**
Ø1844: 72, 74: *Actinia candida*; 72, 74: *Actinia plumosa*; 74: *Metridium plumosum*; 74: *Ectacmaea candida*.

***Pennatula phosphorea* Linnaeus, 1758**
Ø1844: 77, 81, 83, 84: *Pennatula phosphorea*.

***Sagartiogeton viduatus* (Müller, 1776)**
Ø1844: 74: *Actinia viduata*; 74: *Isacmaea viduata*.

***Stomphia coccinea* (Müller, 1776)**
Ø1844: 72, 74: *Actinia coccinea*; 74: *Isacmaea coccinea*.

***Urticina felina* (Linnaeus, 1761)**
Ø1844: 72: *Actinia holsatica*.

***Virgularia mirabilis* (Müller, 1776)**
Ø1844: *Virgularia mirabilis*.

Cnidaria, Hydrozoa.

***Abietinaria abietina* (Linnaeus, 1758)**
Ø1844: 74: *Sertularia abietina*.

***Clava multicornis* (Forsskål, 1775)**
Ø1844: 71: *Coryne squamata*.

***Hydractinia echinata* (Fleming, 1828)**
Ø1844: 74: *Alcyonidium echinatum*.

***Hydrallmania falcata* (Linnaeus, 1758)**
Ø1844: 74: *Plumularia falcata*.

***Kirchenpaueria pinnata* (Linnaeus, 1758)**
Ø1844: 74: *Plumularia pinnata*.

***Obelia geniculata* (Linnaeus, 1758)**
Ø1844: 65, 71: *Campanularia geniculata*.

Cnidaria, Scyphozoa.

***Aurelia aurita* (Linnaeus, 1758)**
Ø1844: 67: *Medusa aurita.*

***Lucernaria quadricornis* Müller, 1776**
Ø1844: 65, 71: *Lucernaria quadricornis.*

Acoela.

***Convoluta convoluta* (Abildgaard, 1806)**
Ø1844: *Convoluta paradoxa.*

Platyhelminthes.

***Niobe zonata* Girard, 1852**
Ø1844: 70: *Planaria limacina.*

***Vortex mytili* Ørsted, 1843**
Ø1844: 69: *Telostoma Mytili.*

***Bothrioplana semperi* Braun, 1881**
Ø1844: 83: *Planaria lactea.*

***Cryptocelides loveni* Bergendal, 1890**
Ø1844: 79: *Typhlolepta coeca.*

***Leptoplana nigripunctata* Ørsted, 1844**
Ø1844: 79: *Leptoplana nigripunctata.*

***Notoplana atomata* (Müller, 1776)**
Ø1844: 79: *Leptoplana atomata.*

***Archilopsis unipunctata* (Fabricius, 1826)**
Ø1844: 69: *Monocelis unipunctata.*

***Monocelis lineata* (OF Müller, 1773) Oersted, 1843**
Ø1844: 69, 83: *Monocelis lineata;* 69: *Monocelis rutilans.*

***Graffiellus croceus* (Fabricius, 1826) Hornung, 2016**
Ø1844: 69: *Prostoma suboviforme.*

***Provortex littoralis* (Ørsted, 1843) Graff, 1882** (uncertain)
Ø1844: 69: *Vortex littoralis.* According to Graff (1882: p.348) the original description is inadequate. It could also be a species of *Plagiostomum.*

***Typhloplana marina* Ørsted, 1843** (nomen dubium)
Ø1844: 69: *Typhloplana marina.*

***Dendrocoelum lacteum* (Müller, 1774)**
Ø1844: 68: *Dendrocoelum lacteum.*

***Foviella affinis* (Ørsted, 1843)**
Ø1844: 69: *Planaria affinis.*

***Planaria torva* (OF Müller, 1773)**
Ø1844: 68, 83: *Planaria torva.*

***Procerodes littoralis* (Strøm, 1768)**
Ø1844: 64, 68, 83: *Planaria Ulvae.*

Nemertea, Heteronemertea.

***Lineus longissimus* (Gunnerus, 1770)**
Ø1844: *Gordius marinus.* Synonym according to Gibson (1995).

***Lineus viridis* (Müller, 1774)**
Ø1844: 66, 69, 83: *Nemertes olivacea.*

Nemertea, Hoplonemertea.

***Amphiporus lactifloreus* (Johnston, 1828)**
Ø1844: 80: *Polystemma roseum.*
Uncertain species. See Gibson (1995) for references.

***Amphiporus pellucidus* (Ørsted, 1843)**
Ø1844: 80: *Polystemma pellicudum.*
Uncertain species. See Gibson (1995) for references.

***Gibsonnemertes spectabilis* (Quatrefages, 1846)**
Ø1844: 69: *Nemertes badia.*

***Oerstedia dorsalis* (Abildgaard, 1806)**
Ø1844: 79: *Tetrastemma varicolor*
Synonym according to Gibson (1995).

***Nemertes flaccida* Ørsted, 1843**
Ø1844: *Nemertes flaccida.*
Uncertain species. See Gibson (1995) for references.

***Nemertes maculata* Ørsted, 1843**
Ø1844: *Nemertes maculata.*
Uncertain species. See Gibson (1995) for references.

***Nemertes pusilla* Ørsted, 1843**
Ø1844: 80: *Nemertes pusilla.*
Uncertain species. See Gibson (1995) for references.

***Nipponnemertes pulchra* (Johnston, 1837)**
Ø1844: 69: *Polystemma pulchrum.*

***Oerstedia dorsalis* (Abildgaard, 1806)**
Ø1844: 79: *Tetrastemma fuscum.*
Synonym according to McCaul (1963) and Gibson (1995).

***Tetrastemma assimile* Ørsted, 1844**
Ø1844: 69: *Tetrastemma assimile.*

***Tetrastemma bioculatum* Ørsted, 1843**
Ø1844: 69: *Tetrastemma bioculatum.*

***Tetrastemma melanocephalum* (Johnston, 1837)**
Ø1844: 66, 69: *Nemertes melanocephala.*

***Tetrastemma subpellucidum* Ørsted, 1843**
Ø1844: 69: *Tetrastemma subpellucidum.*

Nemertea, Palaeonemertea.

***Cephalothrix linearis* (Rathke, 1799)**
Ø1844: 69: *Cephalothrix coeca*; 79: *Cephalothrix bioculata.*

***Cephalothrix rufifrons* (Johnston, 1837)**
Ø1844: 69: *Memertes bioculata* [= *Nemertes*, misspelling]; 79: *Astemma rufifrons.*

Tubulanus annulatus (Montagu, 1804).
Ø1844: 69: *Gordius littoreus.* Müller [1786) called it "Den ternede traadorm" [the checked filamentous worm] and describes it as "filiformis albus rubro maculatus", thread-shaped white red-spotted. This is probably *Tubulanus annulatus*, a species still occurring in the Sound today. There is, however, also a possibility that it could be *Tubulanus superbus* (Kölliker, 1845), a species with similar color pattern found in the region, but not yet recorded in the Sound.

Mollusca

Ø1844: 63, 69, 73, 85: Mollusca.

Mollusca, Gastropoda, Saccoglossa.

***Limapontia capitata* (Müller, 1774)**
Ø1844: 70: *Limapontia nigra;*

***Elysia viridis* (Montagu, 1804)**
Ø1844: 72, 73: *Actaeon minutum. Actaeon minutum* is a synonym of *Elysia viridis* according to Meyer & Möbius (1865) and Sars (1878).

Mollusca, Polyplacophora

Ø1844: 72: Chitones.

***Boreochiton ruber* (Linnaeus, 1767)**
Ø1844: 74: *Chiton ruber.*

***Lepidochitona cinerea* (Linnaeus, 1767)**
Ø1844: 73: *Chiton cinereus.*

Gastropoda, Nudibranchia

Ø1844: 61: Nudibranchiata; 63, 71, 72, 73, 85, 88: Gymnobranchia.

***Aeolidia papillosa* (Linnaeus, 1761)**
Ø1844: 72, 73: *Eolidia papillosa.*

***Cadlina laevis* (Linnaeus, 1767)**
Ø1844: 73: *Doris obvelata.*

***Doris lacinulata* Gmelin, 1791**
(nomen dubium)
Ø1844: 70: *Tergipes lacinulatus.* Placed in the Official Index of Rejected and Invalid Specific Names in Zoology by ICZN (Opinion 773). Name No. 863.

***Doris pseudoargus* Rapp, 1827**
Ø1844: *Doris tuberculata.*

***Doris verrucosa* Linnaeus, 1758**
Ø1844: 73: *Doris verrucosa.*

aff. *Doris verrucosa* Linnaeus, 1758
Ø1844: 73: *Doris* nov. sp?

Hero formosa **(Lovén, 1844)**
Ø1844: 73: *Tritonia velata.*
Hero formosa fits with Ørsted's description of *Tritonia velata.* According to Jensen & Knudsen, this species has not been recorded in Denmark, but it exists along the coasts of Norway (Evertsen & Bakken 2005, 2013). Sven Lovén also recorded this species in Sweden at the same time as Ørsted.

Okenia aspersa **(Alder & Hancock, 1845)**
Ø1844: 73: *Idalla caudata.*
Idalla Ørsted, 1844, De regionibus marinis: 73. This name is suppressed under the plenary powers for the purposes of the Principle of Priority but not for the Principle of Homonymy.

Polycera quadrilineata **(Müller, 1776)**
Ø1844: 72, 73: *Polycera quadrilineata.*

Mollusca, Gastropoda

Ø1844: 69, 80: Gastropoda.

Aporrhais pespelecani **(Linnaeus, 1758)**
Ø1844: 76: *Rostellaria pes pelecani*; 80: *Rostellaria pespelecani*; 83: *Rostellaria pespelecani.*

Bittium reticulatum **(da Costa, 1778).**
Cerithium Danicum is a synonym of *Bittium reticulatum* according to Collin (1884).
Ø1844: 70: *Cerithium Danicum.*

Boreotrophon clathratus **(Linnaeus, 1767)**
Ø1844: 80: *Trophon clathratum.*

Buccinidae, whelks
Ø1844: 63, 75, 77, 85, 88: Buccinoidea.

Buccinum undatum **Linnaeus, 1758**
Ø1844: 76, 80: *Buccinum undatum.*

Ecrobia ventrosa **(Montagu, 1803)**
Ø1844: 69: *Paludina Baltica*; 69: *Paludinella Baltica.*
Paludinella baltica (Nilsson, 1824) is described in a footnote by Ørsted and is therefore a validly introduced name and a new synonym. Nilsson (1824) described it as *Paludina balthica.* At that time, *Potamopyrgus antipodarum* (Gray, 1843), a species it could have been confused with, had not been found or recorded in the Baltic.

Eupaludestrina stagnorum **(Gmelin, 1791)**
Ø1844: 69, 83: *Neritina Baltica.*
Not *Theodoxus baeticus* (Lamarck, 1822)

***Lacuna pallidula* (da Costa, 1778)**
Ø1844: 70: *Lacuna pallidula.*

Lacuna vincta (Montagu, 1803)
Ø1844: 70: *Lacuna canalis*; 70: *Lacuna quadrifasciata.*

***Littorina fabalis* (Turton, 1825)**
Ø1844: 65, 70: *Littorina fabalis.*

***Littorina littorea* (Linnaeus, 1758)**
Ø1844: 63, 65, 70: *Littorina littorea.*

***Littorina obtusata* (Linnaeus, 1758)**
Ø1844: 65, 70: *Littorina retusa.*

***Lymnaea stagnalis* (Linnaeus, 1758)**
Ø1844: 69, 83: *Limnaea Baltica.*

***Melarhaphe neritoides* (Linnaeus, 1758)**
Ø1844: 70: *Littorina petraea.*

***Neptunea* Röding, 1798**
Ø1844: 80: *Fusus* sp.?

***Neptunea antiqua* (Linnaeus, 1758)**
Ø1844: 76, 80: *Fusus antiquus.*

***Nucella lapillus* (Linnaeus, 1758)**
Ø1844: 61, 70: *Purpura lapillus.*

***Ocenebra erinaceus* (Linnaeus, 1758)**
This record is dubious.
It could also be *Neptunea.*
Ø1844: 80: *Murex.*

***Patella* Linnaeus, 1758**
Ø1844: 61, 66, 72: *Patella.*
Ørsted probably also included *Tectura* and *Testudinalia.*

***Patella pellucida* Linnaeus, 1758**
Ø1844: 72, 73, 83: *Patella pellucida.*

***Peringia* Paladilhe, 1874**
Ø1844: 83: *Paludinella.*

***Peringia ulvae* (Pennant, 1777)**
Ø1844: 64, 65, 69: *Paludinella Ulvae*; 65, 69: *Paludinella vulgaris.*
Paludinella vulgaris Ørsted, 1844 is described in a footnote and is therefore a validly introduced name and a new synonym.

***Pleurotomoides* Bronn, 1831**
Ø1844: 80: *Defrancia* sp.

***Rissoa* Desmarest, 1814**
Ø1844: 61: Turbines [*Turbo*].
This is probably *Rissoa.*

Trochoidea
Ø1844: 63, 85, 88: Trochoidea
Probably *Steromphala cineraria* (Linnaeus, 1758) and other trochospiral gastropods of similar shapes.

***Steromphala cineraria* (Linnaeus, 1758)**
Ø1844: 70: *Trochus cinerarius.*

***Tectura virginea* (Müller, 1776)**
Ø1844: 73: *Patella virginea.*

***Testudinalia testudinalis* (Müller, 1776)**
Ø1844: 73: *Patella tessulata.*

***Theodoxus fluviatilis* (Linnaeus, 1758)**
Ø1844: 70: *Neritina fluviatilis.*

***Tritia reticulata* (Linnaeus, 1758)**
Ø1844: 65, 66, 70, 90: *Nassa reticulata.*

***Turritella ungulina* (Linnaeus, 1758)**
Ø1844: 76, 83: *Turritella ungulina.*

***Velutina velutina* (Müller, 1776)**
Ø1844: 80: *Velutina capuloides.*

Mollusca, Gastropoda, Cephalaspidea.

***Philine aperta* (Linnaeus, 1767)**
Ø1844: 66, 70: *Bulla aperta.*

Mollusca, Gastropoda, Heterobranchia.

***Akera bullata* Müller, 1776**
Ø1844: 65, 70: *Akera bullata.*

Mollusca, Bivalvia

Ø1844: 70, 80: Acephala.

***Abra nitida* (Müller, 1776)**
Ø1844: 81: *Abra nitida.*

***Abra tenuis* (Montagu, 1803)**
Ø1844: 81: *Abra tenuis.*

***Acanthocardia echinata* (Linnaeus, 1758)**
Ø1844: 77, 81: *Cardium echinatum.*

***Aequipecten opercularis* (Linnaeus, 1758)**
Ø1844: 76, 80, 83: *Pecten opercularis.*

***Anomia* sp. (uncertain)**
Ø1844: 80: *Anomia* nov. sp.?

***Arctica islandica* (Linnaeus, 1767)**
Ø1844: 77, 81, 83, 84, 90: *Cyprina Islandica.*

***Astarte striata* Sowerby, 1839**
Ø1844: 81: *Astarte striata.* Høpner Petersen (2001) placed it in the *Astarte montagui* species complex. *Astarte montagui* (Dillwyn, 1817) is common in the Sound and has a non-pelagic larval development (Thorson 1946). Species with this reproductive type usually develop into numerous local forms or species, with a fairly limited geographical distribution. This phenomenon is manifested in Høpner Petersen's monograph (2001). *Nicania striata* Leach in Ross, 1819 is a possible synonym.

***Astarte sulcata* (da Costa, 1778)?**
Ø1844: 81, 83, 84: *Astarte Danmoniensis.*

***Bosemprella incarnata* (Linnaeus, 1758)**
Ø1844: 77, 81: *Tellina deprassa.*

***Cerastoderma edule* (Linnaeus, 1758)**
Ø1844: 19, 70, 84, 90: *Cardium edule.*

***Chamelea gallina* (Linnaeus, 1758)**
Ø1844: 81: *Venus gallina.*

Corbula gibba (Olivi, 1792)
Ø1844: 66, 70: *Corbula nucleus.*

Gari fervensis (Gmelin, 1791)
Ø1844: 81, 83: *Psammobia faeroensis.*

Heteranomia squamula (Linnaeus, 1758)
Ø1844: 80: *Anomia aculeata*; 80: *Anomia squamula.*

Hiatella arctica (Linnaeus, 1767)
Ø1844: 77, 81: *Hiatella arctica.*

Limecola balthica (Linnaeus, 1758)
Ø1844: 65, 70: *Tellina Baltica.*

Lucinoma borealis (Linnaeus, 1767)
Ø1844: *Lucina radula.*
Venus borealis sensu Hanley (1855).

Macomangulus tenuis (da Costa, 1778)
Ø1844: 70: *Tellina tenuis.*

Modiolus modiolus (Linnaeus, 1758)
Ø1844: 61: *Mytilus modiolus*; 76, 81, 83: *Mytilus Modiolus*; 81: *Mytilus Modiolus.*

Musculus subpictus (Cantraine, 1835)
Ø1844: 81, 83: *Modiola discrepans.*

Mya Linnaeus, 1758
Ø1844: 65: *Mya.*

Mya arenaria Linnaeus, 1758
Ø1844: 64, 65, 70, 84: *Mya arenaria.*

Mya truncata Linnaeus, 1758
Ø1844: 81: *Mya truncata.*

Mytilus edulis Linnaeus, 1758
Ø1844: 61, 65, 66, 70, 90: *Mytilus edulis.*

Nicania striata Leach in Ross, 1819
Ø1844: *Astarte striata.*
Probably a local species in the *Astarte montagui* complex (Høpner Petersen 2001).

Nucula Lamarck, 1799
Ø1844: 80: *Nucula* nov. sp.?

Nucula nucleus (Linnaeus, 1758)
Ø1844: 80, 83: *Nucula margaritacea.*

Nuculana minuta (Müller, 1776)
Ø1844: 76, 80, 83: *Nucula rostrata.*
This is a synonym sensu Jeffreys (1863).

Nuculana pernula (Müller, 1779)
Ø1844: 80: *Leda rostrata.*

Nuculana sp. dub.
Ø1844: 80: *Leda intermedia.*

Palliolum striatum (Müller, 1776)
Ø1844: 80, 83: *Pecten striatus.*

Phaxas pellucidus (Pennant, 1777)
Ø1844: 77, 81: *Solen pellucidus.*

Pseudamussium peslutrae (Linnaeus, 1771)
Ø1844: 80, 83: *Pecten septemradiatus.*

Spisula Gray, 1837
Ø1844: 70, 84: *Mactra* nov. spec.?

Spisula solida (Linnaeus, 1758)
Ø1844: 70: *Mactra solida.*

Thyasira flexuosa (Montagu, 1803)
Ø1844: *Cryptodon flexuosum.*

Mollusca, Scaphopoda.

Antalis entalis (Linnaeus, 1758)
Ø1844: 76, 80, 83: *Dentalium entails.*

Sipuncula.

Phascolion (Phascolion) strombus strombus (Montagu, 1804)
Ø1844: 80: *Phascolosoma concharum*

Phascolosoma (Phascolosoma) granulatum Leuckart, 1828
Ø1844: 80: *Phascolosoma granulatum.*

Annelida

Ø1844: 61, 68, 78: Annulata.

Annelida, Canalipalpata.

Amphitrite cirrata Müller, 1776
Ø1844: 79: *Terebella cirrata.*

Cirratulus cirratus (Müller, 1776)
Ø1844: 78: *Cirratulus borealis.*

Dipolydora coeca (Ørsted, 1843)
Ø1844: 78: *Leucodorum coecum.*

Dodecaceria concharum Ørsted, 1843
Ø1844: 78: *Dodecaceria concharum.*

Fabricia stellaris (Müller, 1774)
Ø1844: 68: *Tubularia Fabricia.*

Flabelligera affinis M. Sars, 1829
Ø1844: 79: *Chloraema Edwardsii.*

Harmothoe assimilis (Ørsted, 1843)
Ø1844: 78: *Lepidonote assimilis.*

Harmothoe imbricata (Linnaeus, 1767)
Ø1844: 68: *Lepidonote cirrata.*

Harmothoe impar (Johnston, 1839)
Ø1844: 68: *Lepidonote impar.*

Lepidonotus laevis Ørsted (nomen nudum)
Ø1844: 78: *Lepidonote laevis.*

Lepidonotus squamatus (Linnaeus, 1758)
Ø1844: 76, 78: *Lepidonote punctata.*

Nicolea zostericola Ørsted, 1844
Ø1844: 64, 65, 68: *Terebella zostericola.*

Pherusa plumosa (Müller, 1776)
Ø1844: 79: *Pherusa plumosa.*

Polydora ciliata (Johnston, 1838)
Ø1844: 68: *Leucodorum ciliatum.*

Sabella **Linnaeus, 1767**
Ø1844: 79: *Sabella* sp. nov.?

Sabella pavonina **Savigny, 1822**
Ø1844: 79, 83: *Sabella pavonia.*

Sabellaria **Lamarck, 1818**
Ø1844: 79: *Hermella* sp. nov.?

Spio **Fabricius, 1785**
Ø1844: 65: *Spio.*

Spio filicornis **(Müller, 1776)**
Ø1844: 68: *Spio filicornis.*

Spio seticornis **(Linnaeus, 1767)**
Ø1844: 65, 68, 83: *Spio seticornis.*

Terebellides **M. Sars, 1835**
Ø1844: 79: *Terebellides* sp. nov.

Lumbricillus lineatus **(Müller, 1774)**
Ø1844: 65, 68: *Lumbricillus linearis*; 66, 68: *Lumbricillus verrucosus.*

Mesopachys marina **Ørsted, 1844** (uncertain)
Ø1844: 79: *Mesopachys marina.*

Nais elinguis **Müller, 1774**
Ø1844: 68: *Nais elinguis.*

Paranais litoralis **(Müller, 1784)**
Ø1844: 68: *Nais littoralis.*

Tubifex serpentinus **Ørsted, 1844** (uncertain)
Ø1844: 68: *Tubifex serpentinus.*

Annelida, Dinophilidae.

Dinophilus vorticoides **Schmidt, 1848**
Ø1844: 69: *Vortex capitata.*

Annelida, Eunicida.

Scoletoma fragilis **(Müller, 1776)**
Ø1844: 76, 78: *Lumbrineris fragilis*; 83: *Lumbricus fragilis.*

Annelida, Oligochaeta.

Ichtyobdella sanguinea **Ørsted, 1844**
Ø1844: *Ichtyobdella sanguinea*
Probably a misspelling for *Ichthyobdella*. Uncertain species.

Annelida, Oweniidae.

Galathowenia oculata **(Zachs, 1923)**
Ø1844: 79: *Clymenia tenuissima.*
Arwidsson (1906) wrote: Es mag hier erwähnt werden, daß *Clymenia tenuissima* Ørsted (9, p. 79) offenbar der Gattung *Myriochele* Malmgren angehört, vgl. im übrigen 27, p. 186. [It may be mentioned here that *Clymenia tenuissima* Ørsted (9, p. 79) apparently belongs to the genus *Myriochele* Malmgren, cf. also 27, p. 186.]. *Galathowenia oculata* is the only species in this genus that lives in Danish waters (Kirkegaard 1996).

Annelida, Phyllodocida.

***Aphrodita aculeata* Linnaeus, 1758**
Ø1844: 76, 78: *Aphrodita aculeata.*

***Eteone flava* (Fabricius, 1780)**
Ø1844: 78: *Eteone Sarsii.*

***Eteone maculata* Ørsted, 1843**
Ø1844: 78: *Eteone maculata.*

***Eteone pusilla* Ørsted, 1843**
Ø1844: 78: *Eteone pusilla.*

***Eulalia viridis* (Linnaeus, 1767)**
Ø1844: 78: *Eulalia viridis.*

***Eumida sanguinea* (Ørsted, 1843)**
Ø1844: 78: *Eulalia sanguinea.*

***Glycera alba* (Müller, 1776)**
Ø1844: 76: *Glycera alba.*

***Goniada maculata* Ørsted, 1843**
Ø1844: 78: *Goniada maculata.*

***Halithea hystrix* Lamarck, 1818**
Ø1844: 78: *Aphrodita Hystrix.*

***Hediste diversicolor* (Müller, 1776)**
Ø1844: 64, 68, 83: *Nereis diversicolor.*

***Nephtys assimilis* Ørsted, 1843**
1844: 76, 78: *Nephtys assimilis.*

***Nephtys ciliata* (Müller, 1788)**
Ø1844: 76, 78, 83: *Nephtys borealis.*

***Nereimyra punctata* (Müller, 1788)**
Ø1844: 78: *Castalia punctata.*

***Nereis pelagica* Linnaeus, 1758**
Ø1844: 76, 78, 83: *Nereis pelagica.*

***Pholoe baltica* Ørsted, 1843**
Ø1844: 68: *Pholoe Baltica.*

***Phyllodoce assimilis* Ørsted, 1843** (nomen dubium)
Ø1844: 68: *Phyllodoce assimilis.*

***Phyllodoce groenlandica* Ørsted, 1842**
Ø1844: 78: *Phyllodoce Groenlandica.*

***Phyllodoce mucosa* Ørsted, 1843**
Ø1844: 68: *Phyllodoce mucosa.*

***Platynereis dumerilii* (Audouin & Milne Edwards, 1833)**
Ø1844: 64, 68: *Heteronereis fucicola*; 68: *Nereilepas variabilis*; 64, 68: *Nereis zostericola.*

***Syllis armillaris* (Müller, 1776)**
Ø1899: 78: *Syllis armillaris.*

Annelida, Sabellida.

***Fabricia stellaris* (Müller, 1774)**
1844: 68: *Amphicora Sabella.*

***Spirorbis (Spirorbis) spirorbis* (Linnaeus, 1758)**
Ø1844: 68: *Spirorbis nautiloides.*

***Arenicola* Lamarck, 1801**
Ø1844: 65: *Arenicola.*

***Arenicola marina* (Linnaeus, 1758)**
Ø1844: 79: *Lumbricus marinus*; 64, 65, 68: *Arenicola piscatorum.*

***Clymene amphistoma* Savigny in Lamarck, 1818** (taxon inquirendum)
Ø1844: 79: *Clymene amphistoma.*

***Clymene ebiensis* Milne Edwards, 1843** (nomen dubium)
Ø1844: 79: *Clymenam Ebiense.*

***Ophelina acuminata* Ørsted, 1843**
Ø1844: 78: *Ophelina acuminata.*

***Polyphysia crassa* (Ørsted, 1843)**
Ø1844: 78: *Eumenia crassa.*

***Praxillella praetermissa* (Malmgren, 1865)**
Ø1844: 79: *Clymene intermedia.*

***Scoloplos armiger* (Müller, 1776)**
Ø1844: 76, 78, 83: *Scoloplos armiger.*

***Nais equisetina* Duges, 1837** (taxon inquirendum)
Ø1844: 68: *Nais equisetina.*
This is an indeterminable opheliid.

***Travisia forbesii* Johnston, 1840**
Ø1844: 78, 83: 78, 83: *Ophelia mamillata.*

***Chaetopterus variopedatus* (Renier, 1804).**
Ø1844: 78: *Chaetopterus norvegus.*
Possibly *Chaetopterus norvegicus* M. Sars, 1835

***Amphicteis gunneri* (M. Sars, 1835)**
Ø1844: 79: *Amphitrite Gunneri.*

***Amphictene auricoma* (Müller, 1776)**
Ø1844: 66: *Amphitrite auricoma;* 68: *Amphictene auricoma.*

Entoprocta.

***Pedicellina cernua* (Pallas, 1774)**
Ø1844: 82: *Pedicellina echinata.*

***Barentsia gracilis* (Sars, 1835)**
Ø1844: 82: *Pedicellina gracilis.*

Bryozoa.

***Alcyonidium gelatinosum* (Linnaeus, 1761).**
Ø1844: 74: *Alcyonidium gelatinosum.*
Ørsted refers to Johnstone and Johnstone refers to Pallas.

***Bugulina avicularia* (Linnaeus, 1758)**
Ø1844: 74: *Cellaria avicularia.*

***Caberea ellisii* (Fleming, 1814)**
Ø1844: 74: *Flustra setacea.*

***Crisia eburnea* (Linnaeus, 1758)**
Ø1844: 74: *Crisia eburnea.*

***Electra pilosa* (Linnaeus, 1767)**
Ø1844: 74: *Flustra Pilosa.*

***Flustra foliacea* (Linnaeus, 1758)**
Ø1844: 74: *Flustra foliacea.*

***Membranipora membranacea* (Linnaeus, 1767)**
Ø1844: 65, 71: *Flustra membranacea.*

Scrupocellaria scruposa (Linnaeus, 1758)?
Ø1844: 74: *Cellaria* sp. nov.?
In between *Bugulina avicularia* and *Scrupocellaria scruposa*, according to Ørsted. It is probably the latter.

Scrupocellaria scruposa (Linnaeus, 1758)
Ø1844: 74: *Cellaria scruposa.*

Tubulipora liliacea (Pallas, 1766)
Ø1844: 74: *Tubulipora transversa.*

Nematoda.

Anguillula oculata Ørsted, 1844
Ø1844: 80: *Anguillula oculata.*
Species dubium according to Diesing (1861: 626).

Enchelidium marinum Ehrenberg, 1836
Ø1844: 69: *Enchelidium marinum.*

Enoplus oculatus (Ørsted, 1844) Diesing, 1861 (species inquirendum)
Ø1844: 80: *Angvillula oculata.*

Pontonema muelleri Diesing, 1861
Ø1844: 69: *Vibrio marina*;
80: *Angvillula marina.*
Diesing (1861: 623) designated *Pontonema muelleri* as a substitute name for *Vibrio Anguillula marinus* Müller, 1783, a synonym of *Vibrio marinus* and *Anguillula marina* Ørsted, 1844 (Gerlach & Riemann 1974: 599).

Crustacea

Ø1844: 67, 73, 77, 84: Crustacea.

Crustacea, Decapoda

Ø1844: 76: Crustacea Decapoda.

Carcinus maenas (Linnaeus, 1758)
Ø1844: 65, 67: *Carcinus Moenas.*

Crangon crangon (Linnaeus, 1758)
Ø1844: 64, 67: *Crangon vulgaris.*

Galathea strigosa (Linnaeus, 1761)
Ø1844: 77: *Galathea strigosa.*

Geryon trispinosus (Herbst, 1803)
Ø1844: 67: *Geryon tridens.*

Hippolyte gaimardii H. Milne Edwards, 1837
Ø1844: 67: *Hippolyte Gaimardii.*

Homarus gammarus (Linnaeus, 1758)
Ø1844: 77: *Homraus vulgaris*, typographic error for *Homarus.*

Hyas araneus (Linnaeus, 1758)
Ø1844: 76, 77: *Hyas Araneus.*

Inachus phalangium (Fabricius, 1775)
Ø1844: 76, *Stenorynchus phalangium*; 77: *Stenorynchus Phalangium.*

Lithodes maja (Linnaeus, 1758)
Ø1844: 67: *Lithodes arctica.*

Pagurus bernhardus (Linnaeus, 1758)
Ø1844: 76, 77, 83: *Pagurus Bernhardus.*

Palaemon adspersus Rathke, 1837
Ø1844: 64, 66, 67: *Palaemon Squilla.*

Pinnotheres pisum (Linnaeus, 1767)
Ø1844: 67: *Pinnotheres Pisum.*

Crustacea, Mysida.

Praunus flexuosus (Müller, 1776)
Ø1844: 64, 67: *Mysis flexuosus.*

Crustacea, Cumacea.

Diastylis lucifera (Krøyer, 1837)
Ø1844: 78: *Cuma lucifera.*

Diastylis rathkei (Krøyer, 1841)
Ø1844: 78: *Cuma Rathkii.*

Leucon nasica (Krøyer, 1841)
Ø1844: 78: *Cuma nasica.*

Crustacea, Tanaidacea.

Heterotanais oerstedii (Krøyer, 1842)
Ø1844: 67: *Tanais Curculio,* 67: *Tanais Ørstedii.*

Crustacea, Isopoda.

Cyathura carinata (Krøyer 1847)
Ø1844: 67: *Anthura* (*arctica*?). Ørsted uses this name and refers to Krøyer (Kr.), but the species was described by the latter author 3 years later.

Idotea balthica (Pallas, 1772)
Ø1844: 78: *Asellus marinus;* 64, 67: *Idotea tricuspidata.*

Idotea emarginata (Fabricius, 1793)
Ø1844: 67: *Idotea emarginata.*

Idotea pelagica Leach, 1816
Ø1844: 67: *Idotea pelagica.*

Jaera (Jaera) albifrons Leach, 1814
Ø1844: 64, 67, 83: *Jaera albifrons.*

Sphaeroma Bosc, 1801
Ø1844: 67: *Sphaeroma* sp. nov?

Crustacea, Amphipoda.

Ampithoe Leach, 1814
Ø1844: 78: *Amphitoe* sp. nov?

Caprellidae Leach, 1814
Ø1844: 61, 72: Caprellidae.

Caprella Lamarck, 1801
Ø1844: 72: *Caprella.*

Corophium volutator (Pallas, 1766)
Ø1844: 64, 83: *Corophium longicorne.*

Ericthonius punctatus **(Spence Bate, 1857)**
Ø1844: 78: *Podocerus Leachii.*

Gammarus locusta **(Linnaeus, 1758)**
Ø1844: 67: *Gammarus Locusta.*

Gammarus **sp. indet.**
Ø1844: 67: *Gammarus Sabbini.* Uncertain species. For identification of the amphipod species in the region, see Lincoln (1979).

Hyperia **Latreille, 1823**
Ø1844: 67: *Hyperia* sp. nov?

Hyperoche medusarum **(Kröyer, 1838)**
Ø1844: 67: *Metoecus Medusarum.*

Orchestia gammarellus **(Pallas, 1766)**
Ø1844: 64, 67: *Orchestia littorea.*

Phtisica marina **Slabber, 1769**
Ø1844: 73: *Leptomera pedata.*

Talitrus saltator **(Montagu, 1808)**
Ø1844: 64, 67: *Talitrus saltator.*

Crustacea, Cirripedia.

Balanus **Costa, 1778**
Ø1844: 61, 66: *Balanus.*

Balanus balanus **(Linnaeus, 1758)**
Ø1844: 78: *Balanus sulcatus.*

Semibalanus balanoides **(Linnaeus, 1767)**
Ø1844: 64: *Balanus balanoides.*

Arachnida
Ø1844: 67, 84: Arachnida.

Diptera (Insecta)
Ø1844: 83: Diptera.

Pycnogonida

Ø1844: 61, 72: Pycnogonidae.

Anoplodactylus petiolatus **(Krøyer, 1844)**
Ø1844: 73: *Phoxichilidium petiolatum.*

Nymphon grossipes **(Fabricius, 1780)**
Ø1844: 73: *Nymphon grossipes.*

Phoxichilidium **Milne Edwards, 1840**
Ø1844: 73: *Phoxichilidium* nov. sp?

Phoxichilidium femoratum **(Rathke, 1799)**
Ø1844: 73: *Phoxichilidium femoratum.*

Pycnogonum **Brünnich, 1764**
Ø1844: 72: *Pycnogonum.*

Pycnogonum litorale **(Strøm, 1762)**
Ø1844: 73: *Pycnogonum littorale.*

Arthropoda, Acarina.

***Thalassarachna basteri* (Johnston, 1836)** (uncertain)
Ø1844: 67: *Acarus Basteri*; 67: *Acarus setosus.*

Echinodermata

Ø1844: 70, 74, 77, 81: Echinodermata.

Echinodermata, Asteroidea

Ø1844: 61: Asteridae.

***Asterias rubens* Linnaeus, 1758**
Ø1844: 65, 70: *Asteracanthion violaceus*; 70: *Asterias rubra*; 70: *Asterias violacea.*

***Astropecten irregularis* (Pennant, 1777).**
Ø1844: 81: *Asterias arantiaca*; 81: *Astropecten arantiacus.* Not *Astropecten aranciacus* (Linnaeus, 1758) as it is a different species belonging to the Lusitanian (Mediterranean) fauna.

***Crossaster papposus* (Linnaeus, 1767)**
Ø1844: 77, 81: *Solaster papposus*; 81: *Asterias papposa.*

***Solaster endeca* (Linnaeus, 1771)**
Ø1844: 77, 81: *Solaster Endeca*; 80: *Asterias endeca.*

***Stichastrella rosea* (Müller, 1776)**
Ø1844: 81: *Asteracanthion roseus*; 81: *Asterias rosea.*

Echinodermata, Ophiuroidea.

***Amphiura filiformis* (Müller, 1776)**
Ø1844: 77, 81, 84: *Ophiolepis filiformis*; 81: *Asterias filiformis.*

***Ophiopholis aculeata* (Linnaeus, 1767)**
Ø1844: 77, 81: *Ophiolepis scolopendrica*; 81: *Asterias aculeata.*

***Ophiothrix fragilis* (Abildgaard in Müller, 1789)**
Ø1844: 81: *Asterias fragilis*; 81: *Ophiothrix fragilis.*

***Ophiura ophiura* (Linnaeus, 1758)**
Ø1844: 81: *Asterias ophiura*; 81: *Ophiolepis ciliate.*

Echinodermata, Echinoidea.

***Echinocardium flavescens* (Müller, 1776)**
Ø1844: 77: *Spatagus flavescens*; 81: *Spatangus flavescens.*

***Echinocyamus pusillus* (Müller, 1776)**
Ø1844: 81: *Spatangus pusillus.*

***Echinus esculentus* Linnaeus, 1758**
Ø1844: 65, 70, 84: *Echinus esculentus*; 84: *Spatagus esculentus.*

Echinodermata, Holothuroidea

Ø1844: 72: Holothuriae.

***Cucumaria frondosa* (Gunnerus, 1767)**
Ø1844: 74: *Holothuria pentactes*; 74: *Pentacta pentactes*.

***Psolus phantapus* (Strussenfelt, 1765)**
Ø1844: 72, 74: *Holothuria phantopus*; 74, 84: *Psolus phantopus*.

***Psolus squamatus* (Müller, 1776)**
Ø1844: 74: *Cuvieria squamata*; 74: *Holothuria squamata*.

***Thyone fusus* (Müller, 1776)**
Ø1844: 74: *Holothuria fusus*.

Chordata, Ascidiacea

Ø1844: 61, 72: Ascidiae; 70: Tunicata.

***Ciona intestinalis* (Linnaeus, 1767)**
Ø1844: 65: *Ascidia intestinalis*; 70: *Phalusia intestinalis*.

***Halocynthia papillosa* (Linnaeus, 1767)**
Ø1844: *Ascidia rustica*.

***Molgula manhattensis* (De Kay, 1843)**
Ø1844: *Ascidea tubifera*.

Chordata, Teleostei

Ø1844: 84: Pisces.

***Ammodytes tobianus* Linnaeus, 1758**
Ø1844: 66: *Ammodytes Tobianus*.

***Anguilla anguilla* (Linnaeus, 1758)**
Ø1844: 66: *Angvilla vulgaris*.

***Gasterosteus aculeatus* Linnaeus, 1758**
Ø1844: 66: *Gasterosteus aculeatus*.

***Myoxocephalus scorpius* (Linnaeus, 1758)**
Ø1844: 66 *Cottus Scorpius*.

***Pholis gunnellus* (Linnaeus, 1758)**
Ø1844: 66: *Gunellus vulgaris*.

***Platichthys flesus* (Linnaeus, 1758)**
Ø1844: 66: *Platessa Flesus*.

***Spinachia spinachia* (Linnaeus, 1758)**
Ø1844: 66: *Spinachia vulgaris*.

***Zoarces viviparus* (Linnaeus, 1758)**
Ø1844: 66: *Zoarces viviparus*.

References

Arwidsson, I. 1906. *Studien über die Skandinavischen und Arktischen Maldaniden nebst Zusammenstellung der übrigen bisher bekannten Arten dieser Familie*. Inaugural Dissertation, Uppsala.

Carlgren, O. 1945. *Polypdyr (Coelenterata) III. Koraldyr. Danmarks fauna*. Copenhagen: G.E.C. Gads Forlag.

Collin, J. 1884. *Om Limfjordens tidligere og nuværende marine fauna, med særlig hensyn til bløddyrfaunaen*. Copenhagen: Gyldendalske Boghandels Forlag.

Cuvier, G. 1817. *Le Règne Animal Distribué Selon son Organisation, pour Servir de Base à l'Histoire Naturelle des Animaux et d'Introduction à l'Anatomie Comparée*. Paris: Déterville.

Diesing, K.M. 1861. 'Revision der Nematoden.' *Sitzungsberichte der kaiserlichen Akademie der Wissenschaften zu Wien, Mathematisch-Naturwissenchaftlische Classe* 42: 595–763.

Drouet, F. 1968. 'Revision of the Classification of the Oscillatoriaceae.' *The Academy of Natural Sciences of Philadelphia, Monograph* 15: 1–370.

Drouet, F. 1973. *Revision of the Nostacaceae with Cylindrical Trichomes (Formerly Scytonemataceae and Rivulariaceae)*. New York: Hafner Press.

Gerlach, S.A. & Riemann, F. 1974. *The Bremerhaven Checklist of Aquatic Nematodes. A Catalogue of Nematoda Adenophorea Excluding the Dorylaimida*. Part 2. Veröffentlichungen des Instituts für Meeresforschung in Bremerhaven, Supplement 4, 403–736.

Gibson, R. 1995. 'Nemertean genera and species of the world: an annotated checklist of original names and description citations, synonyms, current taxonomic status, habitats and recorded zoogeographic distribution.' *Journal of Natural History* 29: 271–562.

Graff, L. von. 1882. *Monographie der Turbellarien 1. Rhabdocoelida*. Leipzig: Verlag Wilhelm Engelmann.

Guiry, M.D. & Guiry, G.M. 2022. *AlgaeBase*. World-wide electronic publication, National University of Ireland, Galway. https://www.algaebase.org.

Hanley, S. 1855. *Ipsa Linnaei Conchylia*. London: William and Norgate.

Heiberg, P.A.C. 1863. *Conspectus Criticus Diatomacearum Danicarum*. Copenhagen: Wilhelm Priors Forlag.

Høpner Petersen, G. 2001. *Studies on some Arctic and Baltic* Astarte *species (Bivalvia, Mollusca)*. Meddelelser om Grønland, Bioscience 52. Copenhagen: Danish Polar Center.

Jeffreys, J.G. 1863. *British Conchology, or an account of the Mollusca which now inhabit the British Isles and the surrounding Seas. Volume II. Marine Shells, comprising the Brachiopoda, and Conchifera from the family Anomiidae to that of Mactridae.* London: John van Voorst.

Kirkegaard, J.B. 1996. *Havbørsteorme. II. Danmarks fauna* 86. Copenhagen: Dansk Naturhistorisk Forening.

Køie, M. & Kristiansen, A. 2014. *Havets dyr og planter*. Second edition. Copenhagen: Gyldendal.

Liebmann, F.M. 1839. 'Bemaerkninger og tillaeg til den Danske algeflora.' *Naturhistorisk Tidsskrift* 2: 464–94.

Lokhorst, G.M. & Trask, B.J. 1981. 'Taxonomic studies on *Urospora* (Acrosiphoniales, Chlorophyceae) in Western Europe.' *Acta botanica neerlandica* 30(5/6): 353–431.

Nilsson, S. 1824. *Historia Molluscorum Sveciae terrestrium et fluviatilium breviter delineata.* Lund: Litteris Berlingianus.

Thorson, G. 1946. *Reproduction and larval development of Danish marine bottom invertebrates.* Meddelelser fra Kommissionen for Danmarks Fiskeri- og Havundersøgelser. Serie: Plankton 4(1): 1–523.

WoRMS Editorial Board. 2022. *World Register of Marine Species.* Available from https://www.marinespecies.org at VLIZ. doi:10.14284/170.

Appendix

Translation of: Ørsted, A.S. 1849. 'Iagttagelser over en hidtil ukendt almindelig Udbredelse af microskopiske Planter i Verdenshavet.' *Videnskabelige Meddelelser fra den naturhistoriske Forening i Kjöbenhavn for Aarene 1849 og 1850*: 6–11.

Observations on an unprecedented prevalence of microscopic plants in the world ocean.

By Mag. A.S. Ørsted.

When the naturalist leaves his northern home in the early spring months to cross the vast ocean to distant continents, then he will for a few weeks be prevented from satisfying his wish to collect and examine marine organisms by the erratic, often violent, winds and his unfamiliarity with the sea. He therefore seizes with all the greater joy the opportunity which the calm sailing in the trade winds offers to make observations of the sea's animals and plants, which here often in abundant quantity surround the ship.

On the voyage to the West Indies in 1815 [should be 1845], at about the height of Madeira, I was still preoccupied with animals that I had standing in large glasses of seawater, in order to study their way of life and development, and then found that the water was never perfectly clear, but, especially by being held against the light, showed a peculiar opacity, which arose from very fine fluffs which remained floating in the whole mass of water.

I assumed that these fluffs were organic parts of dissolved animals and plants, but by looking at them under the microscope it turned out that they were independent microscopic plants, thus filling the sea in innumerable quantities. Each fluff is composed of very fine oscillatoria-like wires, which either lie next to each other so that they form a bundle, or radiate in all directions from a point. The individual filaments show essentially the same construction that distinguishes the Oscillatoriae [Cyanophyceae], being very thin, transparent and having an indistinct segmentary division; on the other hand, they do not have such a great stiffness that in this respect they are more similar to the Confervae [green algae] related to the Oscillatoriae.

When I assumed that the presence of such a large amount of microscopic plants in the world ocean was not without significance in terms of the explanation of some hitherto enigmatic conditions in the natural history of the sea, if it turned out that they had a large and common distribution, then my attention was still addressed to this throughout the rest of the voyage to the Caribbean, later from here to Central America and also by a voyage on the Pacific Ocean. Several times a day, the seawater was examined to find out if these plants were present in it. It turned out then that they occurred everywhere in larger or smaller quantity. Only once were they found in such densely compacted mounds that they gave the sea surface a brownish color, namely, near San Juan de Nicaragua. They were never completely absent. However, it was not the same species everywhere; but gradually there were 5 different species, which would probably be attributed to 2 or 3 different genera of the family of the Oscillatorians [Cyanophyceae].

Since these microscopic plants occur everywhere in the part of the Atlantic and Pacific Oceans that I have had the opportunity to study, I do not think one will find it hasty to conclude that they are also found in the rest of the world, but that they have hitherto eluded the attention of natural scientists because of their small size, just as they were not previously observed in the part of the ocean where, according to the observations reported here, they have a common distribution. From this I think it is already justified to derive the general result: *the water in the world ocean, even when for immediate consideration it turns out to be perfectly clear, on closer examination is found to contain a large amount of microscopic plants.*[1] However, the validity of this phrase is further corroborated by many experiences of the sporadic appearance of microscopic plants belonging to the same or very close species that have been observed in various places in the world. Almost all natural

---

1. Although they have so far only been observed near the surface, there is little doubt that they do exist, probably in decreasing quantities, at all depths, as, like the free-swimming freshwater plants, they sink to the bottom as they die away.

scientists who have made voyages across the ocean report large stretches of the sea, often of many miles' extension, where the surface of the incredible amount of such plants accumulated in dense crowds caused their own color. (See accounts of Banks and Solander, Peron, Chamisso, Hinds, Darwin, Montagne, etc.). However, there is no place where they appear to be so common and to such an extent as in the South Seas, especially near New Holland [Australia]. It is also only there that they have been given their own name among the seafarers. They call them sea sawdust, as they appear as sawdust scattered across the sea. It is a similar plant, but of a red color, which, according to the observations of Ehrenberg and several French travelers, is found in such quantity in the Red Sea that one must assume that it is the one that gave rise to the name of this sea. Since these microscopic plants, according to the above-mentioned observation, have a general distribution over a large part of the Atlantic Ocean, where they had hitherto not been found sporadically prominent in such quantity that they have given the sea a color of its own, it becomes so much more likely that a careful study will show that they are also found in the southern part of the Atlantic Ocean and in the South Seas, where at certain times and in certain places they have been found in incredible numbers, that they also have a common distribution there, but that so far they have only attracted the attention of travelers where they were found in easily observed masses.

If it is thus assumed that the water in the sea in general, although it also seems perfectly clear, nevertheless contains microscopic plants in abundant quantity, then the disparity which has hitherto had to be assumed between the mass of animals and plants in the sea disappears. While on the one hand it was known that the plants have only a very limited distribution in the sea, as they, as far as they are attached, are only found near the coast up to a depth of 200–300 feet, and insofar as they are free-drifting, are confined to a small part of the Atlantic Ocean (namely the so-called Sargasso Sea, Mar de Sargasso), then it is on the other hand a commonly recognized experience that the animals are widespread throughout the sea, partly as they have been found at the greatest depths that have been studied so far, partly due to the fact that the animals living in the open sea everywhere fill the seas, just as it has also been proven by Ehrenberg's extensive studies that the sea, even when it seems perfectly clear, however, contains a large amount of microscopic animals. But by this lack of plants in the vast majority of the sea, which, however, is filled with animals, it was impossible to explain from whence all these animals ultimately obtained their nourishment; for surely it is true that most of the animals of the sea live on animal food, the larger ones on the smaller ones, etc. But what about the smallest ones? From where do the many infusoria [ciliates] and the many small Entomostraca, which fill the sea to such an extent that it is conspicuous that they, by their luminous property, glorify the nights in the tropical seas, from where do all these smallest animals of the sea obtain their food? In other words, it has hitherto been an unresolved issue of whether one could assume that the same law applies to the animals that live in the sea as to those that live on land, namely that all animal food eventually comes from the plant kingdom, that all the carbon that makes up the largest part of the animal body comes from the plants. There is now little doubt that it is the role of these microscopic plants, which widely occur throughout the world, that they provide the necessary plant material for the nutrition of the smallest animals, that the same law applies in this respect with regard to animal nutrition in the sea as on land: *that also in the sea all animal food eventually comes from the plant kingdom.*

By examining the contents of the intestinal tract of many of the small animals living in the open sea, I have been strengthened in the correctness of this statement, as it was always found to consist of a large proportion of these microscopic plants.

(Since Mr. Prof. Steenstrup a short time ago (see *Oversigt over det kongelige danske Videnskabernes Selskabs Forhandlinger* (1849), p. 52) has drawn attention to some microscopic organisms, which have much in common with the microscopic plants that have been made the subject of consideration here, and since the plant nature of these organisms has been questioned by him, it may not be superfluous to note that, by a sample kindly given to me, I have been able to examine whether they correspond to those organisms, on whose plant nature the above proof is based. The "brown" form is probably very similar to the plants I found, but still of a different genus.[1] The "green", on the other hand, whose nature has not become clear to me, seems to me to be very different. So, while I do not doubt that the latter may belong to the animal kingdom, the structure and the mode of development do not allow me to assume that the same should apply to the brown form.)

A more detailed account of these observations, accompanied by informative figures, will be published elsewhere.

---

1. Its generic conformity or inconsistency with those of Mag. Ørsted's observed Forms I can't determine; but it is certain that as a species it can hardly be distinguished from those forms [species] described by Ehrenberg, Montague and from the seafarers' "Sea-sawdust". J. Stp. [Japetus Steenstrup].

About the authors

TOMAS CEDHAGEN was born in Gothenburg, Sweden, in 1954. He studied chemistry, oceanography, geosciences, informatics, and biology at the University of Gothenburg, and received his doctorate degree and became Associate Professor (Docent) in Zoology. He lectured at Tjärnö Marine Biological Laboratory, Sweden, until 1995, after which he became Associate Professor at the Institute for Biology, Aarhus University from 1995 to 2021, and emeritus after that.

JØRGEN LEDET CHRISTIANSEN was born in Raaby, Denmark, in 1944. He studied classical philology and comparative linguistics at Aarhus University. Since 1966 he has taught Greek and Latin in high school and at the universities of Odense and Aarhus. He has published Danish translations of the Apocrypha and of the writings of various Church Fathers.

Pioneering Marine Ecology in the Øresund
Anders Sandøe Ørsted's *De Regionibus Marinis*

Cover, layout and typesetting:
Carl-H.K. Zakrisson and Tod Alan Spoerl
Cover illustration: Detail from *De Regionibus Marinis. Tab. I. Naturhistorisk Kort over Öresundet*. 1844.
Publishing editor: Henrik Jensen
This book is typeset in Kaius og Walbaum
and printed on 115g Munken Print White 15
Printed by Narayana Press, Denmark

Printed in Denmark 2025
1st edition, 1st impression

ISBN 978 87 7597 515 0 (printed book)
ISBN 978 87 7597 860 1 (e-pdf)
ISBN 978 87 7597 859 5 (epub)

Aarhus University Press
Helsingforsgade 25, DK-8200 Aarhus N
unipress@unipress.au.dk
aarhusuniversitypress.dk

Published with the financial support of
Aarhus University Research Foundation.

The carbon emission of this book is calculated to be 2,0 kg according to ClimateCalc. CC–000159/DK